D0594180

RATIONAL
Mysticism

BOOKS BY JOHN HORGAN

The Undiscovered Mind

The End of Science

Rational Mysticism

RATIONAL Mysticism

DISPATCHES FROM THE BORDER BETWEEN SCIENCE AND SPIRITUALITY

JOHN HORGAN

HOUGHTON MIFFLIN COMPANY
BOSTON NEW YORK 2003

Visit our Web site: www.houghtonmifflinbooks.com.

Library of Congress Cataloging-in-Publication Data

Horgan, John, date.
Rational mysticism : dispatches from the border between science and spirituality / John Horgan.
p. cm.
Includes bibliographical references and index.
ISBN 0-618-06027-8
1. Mysticism. 2. Religion and science. I. Title.

BL625.H67 2003
291.1'75 — dc21 2002032281

Book design by Anne Chalmers
Typefaces: Minion, Scala Sans

Printed in the United States of America

QUM 10 9 8 7 6 5 4 3 2 1

For Suzie, Mac, Skye,
and all wild creatures

CONTENTS

RATIONAL Mysticism

Introduction

Lena's Feather

My wife, Suzie, is known in our hometown as a nurturer of birds. One recent spring a neighbor brought her a crow hatchling he had found in the woods. After failing to find its nest, Suzie decided to raise the crow, which she named Lena. When she first arrived, Lena had blue eyes, as all fledgling crows do, and she could barely walk, let alone fly. A cardboard box in the corner of our living room served as her nest. When Suzie approached with grape slices, moistened dog food pellets, and live mealworms, Lena flung her head back and opened her beak wide. Suzie dropped the morsels into Lena's pink gullet, and Lena gulped them down.

Lena was soon hopping and flapping around the living room like a gangly teen, crashing into chairs and windows, poking through our bric-a-brac. After Suzie took her outside onto our deck, Lena launched herself onto the roof of the house and into nearby trees. She always returned for meals, and each night after dinner Suzie brought her inside for the night, until one evening when Lena vanished into the woods. Suzie was distraught, fearing that a hawk or an owl would kill the adolescent bird. But when Suzie went outside at dawn with a plate of worms and grapes, Lena careened out of the sky and skidded onto the deck, cawing.

That pattern persisted. Lena disappeared at night and returned every morning for food and companionship. Because I am my family's earliest riser, she usually greeted me first. As I sipped coffee in my attic office, caws approached through the skylight above my desk, followed by wingbeats and claws scratching shingles. A moment later, Lena peered down at me through the skylight, cooing. When I went out on the deck later to read the newspaper, she crouched at my feet and yanked on my shoelaces or perched on my shoulder and pecked the paper. I pretended to be annoyed, shooing her away, and to my delight she kept coming back.

Lena loved playing tag with our kids, Mac, who was five then, and Skye, who was four. As they chased her, she bounded on the ground before them, occasionally pirouetting behind them and scooting between their legs, staying just beyond their reach. She was fearless. When Mac and Skye swooped back and forth on swings, she stood near the low point of their trajectory and pecked at their rear ends whooshing by. Lena's first love was Suzie. When Suzie came outside, Lena would hop on her shoulder and nestle against her neck, making noises of affection, as did Suzie.

We spent two magical months in this manner, with this wild creature insinuating herself into our lives. One morning as I sat in my office staring at my computer, I heard a howl of anguish from outside. I ran into the back yard and found Suzie sitting on the ground, wailing, with Lena in her lap. Lena's glossy black form was limp, her blue eyes dim. Blood oozed from her beak. She had been playing tag with Mac and Skye. One of them had collided with Lena, breaking her neck. We buried her on a hillock near our house. Suzie planted daffodils and tulips over her grave.

At the time, I was in the midst of research for this book. The next morning, I was to fly to California to take part in a ceremony that called for ingestion of ayahuasca, a powerful psychedelic substance made from two Amazonian plants. Ayahuasca is an Indian word often translated as "vine of the dead." For centuries, shamans in South America have used ayahuasca to propel themselves into trances, during which they travel to a mystical underworld and commune with spirits. Ayahuasca triggers violent nausea, and its visions can be nightmarish. It has nonetheless recently become a sacrament of sorts for spiritual adventurers around the world.

As I packed for the trip that evening, I felt a melancholy that seemed out of proportion to Lena's death, as upsetting as it had been. This creature's demise, I realized, reminded me how fragile all our lives are. Everyone I love — my wife and children — is doomed, and can be taken from me at any moment. My anticipation of the impending ayahuasca session began mutating into dread. I feared that the vine of the dead would force me into a more direct confrontation with death, and I wasn't sure I felt up to the challenge. The people supervising the ayahuasca session had asked each participant to bring a "sacred object," something of personal significance.

So in my knapsack — along with my tape recorder, pens, notebook, and several books — I put one of Lena's feathers.

Looking for The Answer

I cannot recall exactly when I first learned about the extraordinary way of perceiving, knowing, and being called mysticism. Certainly by the early 1970s, when I was in my late teens, the topic was impossible to avoid. Everyone I knew seemed to be reading *Siddhartha, Be Here Now, The Doors of Perception, The Teachings of Don Juan,* and other mystical texts. Everybody was pursuing mystical epiphanies — satori, kensho, nirvana, samadhi, the opening of the third eye — through Transcendental Meditation, kundalini yoga, LSD, or all of the above.

And why not? Spiritual tomes ancient and modern promised that mysticism is a route not only to ultimate truth — the secret of life, the ground of being — but also to ultimate consolation. The supreme mystical state, sometimes called enlightenment, was touted as a kind of loophole or escape hatch in reality, through which we can wriggle out of our existential plight and attain a supernatural, even divine, freedom and immortality.

Along with millions of others in my generation, I puzzled over esoteric mystical books, and I dabbled in yoga, meditation, and psychedelic drugs. I never dedicated myself to the mystical path, however. Friends who had done so — typically by joining one of the countless guru-led groups that sprang up in the 1960s and 1970s — seemed to have abandoned their rationality and autonomy. Also, the insights I gleaned from my own experiences were too confusing, and sometimes frightening, for me to make good use of them. At a time when I was trying to make something of myself, they were a destabilizing influence.

By the early 1980s, I had decided that science represents our best hope for improving our condition — and for understanding who we are, where we came from, where we're going. Some physicists were seeking a so-called theory of everything, an explanation of the physical universe so encompassing that it might solve the biggest riddle of all: Why is there something rather than nothing? Thrilled by science's ambitions, I became a science writer, and for more than a decade I wrote articles about particle physics, cosmology, complexity theory, and other fields that promised great revelations.

Gradually, I came to the conclusion that science can take us only so far in our quest for understanding. Science will not reveal "the mind of God," as the British physicist (and atheist) Stephen Hawking once promised. Science will never give us The Answer, a theory powerful enough to dispel all mystery from the universe forever. After all, science itself imposes limits on what we can learn through rational, empirical inquiry. I spelled out these conclusions in two books: *The End of Science,* which analyzed science as a whole, and *The Undiscovered Mind,* which focused on fields that address the human mind.

In both books, I briefly considered whether mystical experiences might yield insights into reality that can complement or transcend what we learn through objective investigations. In *The End of Science,* I alluded to a drug-induced episode that had been haunting me since 1981. I kept this section short, because I feared it might repel the scientifically oriented readers for whom my book was intended. The opposite reaction occurred. Many readers — including scientists, philosophers, and other supposed rationalists — wrote to tell me that they found the section on mysticism the most compelling part of the book. Readers related their own mystical episodes, some ecstatic, others disturbing. Like me, these readers seemed to be struggling to reconcile their mystical intuitions with their reason.

That was when I first considered writing a book on mysticism. I wasn't sure that the topic would warrant book-length treatment. As recently as 1990 the psychologist Charles Tart, editor of *Altered States of Consciousness,* a collection of scholarly articles on mysticism and other exotic cognitive conditions, complained that so little research had been done since his book's publication in 1969 that it scarcely needed updating. Attempts to reconcile science and mysticism had apparently not progressed much beyond crude studies of meditators' brain waves and claims of vague correspondences between quantum mechanics and Hindu doctrine.

But I soon found that investigations of mysticism are proceeding along a broad range of scholarly and scientific fronts. During the 1990s ordinary consciousness, once considered beneath the notice of respectable scientists, became a legitimate and increasingly popular object of investigation. Emboldened by this trend, some scientists have begun focusing on exotic states of consciousness, including mystical ones. Researchers are sharing results at conferences such as "Worlds of Consciousness," held in 1999 in Basel, Switzerland, the birthplace of LSD; and in books such as *The Mystical Mind, Zen and the Brain,* and *DMT: The Spirit Molecule.*

Their approaches are eclectic. Andrew Newberg, a radiologist at the University of Pennsylvania, is scanning the brains of meditating Buddhists and praying nuns to pinpoint the neural correlates of mystical experience. The Canadian psychologist Michael Persinger tries to induce religious visions in volunteers by electromagnetically stimulating their brains with a device called the God machine. The Swiss psychiatrist Franz Vollenweider has mapped the neural circuitry underlying blissful and horrific psychedelic trips with positron emission tomography. The findings of researchers like these are invigorating long-standing debates among theologians, philosophers, and other scholars about the meaning of mysticism and its relationship to mainstream science and religion.

This upsurge in scientific and scholarly interest has not brought about consensus on mystical matters. Quite the contrary. Scholars disagree about the causes of mystical experiences, the best means of inducing them, their relation to mental illness and morality, and their metaphysical significance. Some experts maintain that psychology and even physics must be completely revamped to account for mysticism's supernatural implications. Others believe that mainstream, materialistic science is quite adequate to explain mystical phenomena. Similarly, scholars disagree about whether mystical visions affirm or undermine conventional religious faith.

Eventually I decided that the time was right after all for a book on mysticism. Most such books, whether written by philosophers of religion, neurologists, or New Age gurus, hew to a particular theory or theology, such as Zen Buddhism or psychedelic shamanism or evolutionary psychology. My goal was to write a book as wide-ranging, up-to-date, and open-minded as possible. The book would be journalistic, based primarily on face-to-face interviews with leading theologians, philosophers, psychologists, psychiatrists, neuroscientists, and other professional ponderers of mysticism. I would assess their respective findings and conjectures, trying to determine where they converge or diverge, where they make sense or go off the deep end. To provide historical context, I would show how recent mystical studies are both corroborating and advancing beyond inquiries undertaken in the past by scholars such as William James and Aldous Huxley. And I would discuss my personal experiences where relevant.

Mysticism's schisms

Mysticism, the human-potential priestess Jean Houston warned me early on in this project, begins in mist, has an I in the middle, and ends in schism. Debate begins with definition. Mysticism is often defined, in a derogatory sense, as metaphysical obfuscation, or belief in ghosts and other occult phenomena. William James mentioned these meanings in his classic 1902 work *The Varieties of Religious Experience* before offering a definition that is still widely cited.* Mysticism, James proposed, begins with an experience that meets four criteria: It is ineffable — that is, difficult or impossible to convey in ordinary language. It is noetic, meaning that it seems to reveal deep, profound truth. It is transient, rarely lasting for more than an hour or so. And it is a passive state, in which you feel gripped by a force much greater than yourself. Two qualities that James did not include in his formal list but mentioned elsewhere are blissfulness and a sense of union with all things.

In *Cosmic Consciousness*, published at around the same time as *The Varieties of Religious Experience*, the Canadian psychiatrist Richard Bucke described an experience that met all of James's criteria. A carriage was bearing Bucke home from an evening lecture when he was overcome by "immense joyousness accompanied or immediately followed by an illumination quite impossible to describe." The experience lasted only a few moments, but during it Bucke "saw and knew" that "the Cosmos is not dead matter but a living Presence, that the soul of man is immortal, that the universe is so built and ordered that without any peradventure all things work together for the good of each and all, that the foundation principle of the world is what we call love and that the happiness of everyone is in the long run absolutely certain."

But in *The Varieties of Religious Experience*, James made it clear that mystical experiences may not be ineffable, transient, passive, blissful, or

---

* *The Varieties of Religious Experience* still exerts a powerful influence over discussions of religious experience and mysticism. "After 100 Years, William James's 'Varieties,' Maybe Not Flawless, Resonates Nonetheless," declared the headline of an article by *New York Times* religion columnist Peter Steinfels (March 9, 2002, p. B6). Although faulted for overemphasizing the subjective dimensions of spirituality and neglecting social aspects, *Varieties* endures, Steinfels said, because James articulated so eloquently what it is like to be poised "on the cusp between belief and unbelief."

unitive. Some mystics describe their supposedly ineffable visions at great length. They may claim to be gifted not just with transient flashes of insight but with a permanent shift in vision. They may feel not passive but powerful, and the power seems to come from inside rather than outside them. And while some mystics feel a blissful unity with all things, others perceive absolute reality as terrifyingly alien. James called these visions "melancholic" or "diabolical."

Even the quality that James called noetic has been challenged. Certain mystics describe their experience as a form of ecstatic forgetfulness or self-dissolution rather than of knowing. To my mind, however, a sense of absolute knowledge is the *sine qua non* of mystical experiences; this noetic component transforms them into something more than transient sensations. "The mystic vision is not a feeling," declares the religious scholar Huston Smith. It is "a seeing, a knowing." The vision may or may not be ineffable, transient, unitive, or blissful, but it must offer some ultimate insight, however strange, paradoxical, and unlike ordinary knowledge. It must grip us with the certainty that we are seeing "the Way Things Are," as the sociologist and Catholic priest Andrew Greeley once put it.

Estimates of the frequency of mystical experiences vary — not surprisingly, given the variability of definitions. A survey carried out in the 1970s found that 33 percent of adult Americans have had at least one experience in which they sensed "a powerful spiritual force that seemed to lift you outside of yourself." A British poll determined that a similar percentage of people have been "aware of, or influenced by, a presence of power." The experiences may be induced deliberately by drugs, meditation, prayer, or other spiritual practices, but they may also be spontaneous responses to natural beauty, music, childbirth, lovemaking, life-threatening events, intense grief, and illness.

Some researchers contend that full-blown mystical experiences are much less common than these surveys indicate. The neurologist and Zen Buddhist James Austin, author of *Zen and the Brain*, suspects that the state he calls absorption — known as samadhi by Hindus and satori by Buddhists — is quite rare. During this state, the external world and one's own self seem to dissolve into a formless unity. Even rarer than absorption, according to Austin, is nirvana, realization, liberation, awakening, enlightenment, in which sporadic flashes of insight yield to a long-term shift in vision.

However rare mystical transcendence is, multitudes are pursuing it. Enlightenment is the telos of the great Eastern religions, Hinduism and Buddhism. Mysticism has played a smaller but still vital role in the history of Judaism, Christianity, and Islam. Countless modern sects, such as the Transcendental Meditation and Hare Krishna movements, also hold out the promise of nirvana to devotees. Centers of "holistic learning," including the Omega Institute in upstate New York, Colorado's Naropa Institute, and the California Institute for Integral Studies, offer courses in what could be called mystical technologies, including Vipassana meditation, shamanic drumbeating, tantric yoga, Kabala studies, and Sufi dancing.

Others seek mystical insights by ingesting psychedelic substances such as LSD and psilocybin mushrooms. Two hundred and fifty thousand Indians who belong to the Native American Church consume the fruit of the peyote cactus as a sacrament. Ayahuasca serves a similar purpose for thousands of members of two fast-growing sects in Brazil — and for a growing number of North Americans and Europeans. Clearly, in the so-called age of science, many of us still look to mysticism for truth and consolation.

But can mystical spirituality be reconciled with science and, more broadly, with reason? To paraphrase the mystical philosopher Ken Wilber, is the East's version of enlightenment compatible with that of the West? If so, what sort of truth would a rational mysticism give us? What sort of consolation? These, I believe, are the most important issues confronting mystical scholars and the millions who are following mystical paths.

While attempting to resolve these basic issues, I will touch on many other questions that motivate today's mystical inquiries: What can neuroscience, psychiatry, and other mind-related fields tell us about the causes of mystical states? Are there any risks in following the mystical path, whether by meditating or ingesting peyote? What is the link between mysticism, madness, and morality? Does belief in mysticism always go hand in hand with belief in parapsychology? What is the nature of the supreme mystical state, sometimes called enlightenment? Will science ever produce a mystical technology powerful enough to deliver enlightenment on demand?

Seeking mystical experts

Mysticism derives from the Greek root *mu*, which means silent or mute. In ancient Greece, the adjective *mystikos* referred to secrets revealed only to those initiated into esoteric sects; mystical knowledge was that which *should* not be revealed. Over time, mystical knowledge came to be defined as that which transcends language and so *cannot* be revealed.* An aphorism from the ancient mystical text the *Tao Te Ching* can be read both ways: "Those who know, do not speak. Those who speak, do not know." In other words, no one who talks about mysticism — including, presumably, the author of the *Tao Te Ching* — really knows anything. Niels Bohr's quip about quantum mechanics comes to mind: Anyone who says he understands quantum mechanics, the great physicist remarked, doesn't know the first thing about it.

Some are nonetheless more qualified than others to talk about quantum mechanics, and the same is true of mysticism. The bulk of this book consists of profiles of those who might be called mystical experts (although that phrase does have an oxymoronic ring). "No ideas but in things," the poet William Carlos Williams once wrote. I suppose my journalistic credo might be "No ideas but in people." In writing about science, I have tried to show that certain theories are best understood not as discoveries plucked whole from some Platonic ether but as embodiments of the aspirations and anxieties of living, breathing individuals. This principle applies at least as much to mystical doctrines such as gnosticism, negative theology, and Zen as it does to superstring theory and psychoanalysis. Scientists' personalities can influence their scientific products, but when it comes to spirituality, personality *is* the product, at least in principle.

Mystical enthusiasts often declare that you cannot comprehend mystical experiences if you have never had one. This attitude smacks of elitism — it recalls Freudians' self-serving claim that only those who have undergone psychoanalysis are qualified to judge it — but there is some truth to it. As the neuroscientist Francisco Varela has said, comprehending mysticism and indeed all aspects of the mind requires both first-person

---

*"There is no really adequate history of the originally Greek qualifier *mystikos* and its various derivatives," the religious scholar Bernard McGinn wrote in *The Foundations of Mysticism* (Crossroad, New York, 1991, p. 344). McGinn emphasized that the term *mysticism* and its modern meaning are relatively recent scholarly inventions. French philosophers apparently started using the term *la mystique* — which English scholars translated into *mysticism* — in the seventeenth century to describe a distinctive set of religious practices, beliefs, and experiences. "No mystics (at least before the present century) believed in or practiced 'mysticism,'" McGinn said (p. xvi). "They believed in or practiced Christianity (or Judaism, or Islam, or Hinduism)."

and third-person perspectives. Hence most, though certainly not all, of my profile subjects claim to have both subjective and objective knowledge of the mystical realm; they can discuss it from the inside and the outside. They also share the belief that mysticism has much to offer us.

There are no clear-cut criteria for judging spiritual expertise. The psychologist Howard Gardner, author of the multiple-intelligence theory of human nature, has made this point. Reasonable standards exist for evaluating scientific, mathematical, athletic, artistic, literary, and musical achievement, Gardner noted, but there is no objective measure for "the attainment of a state of spiritual truth." Some experts I interviewed struck me as wise or "spiritual," but what I looked for primarily was a serious, sustained effort to comprehend mysticism in all its complexity. I also sought experts with diverse perspectives, hoping that illumination might emerge through polyangulation.

Ultimately, the stakes involved in any inquiry into mysticism are philosophical and theological in nature. Hence, this book begins by examining an ongoing debate among philosophers and theologians over mysticism's meaning. Perhaps the most significant issue concerns whether mystical experiences transcend space and time or are all colored to some extent by the mystic's personality and cultural indoctrination. In other words, can a nineteen-year-old engineering student who has taken LSD at a rave discover the same truth and even have the same experience as a sixteenth-century nun in the throes of an epileptic seizure?

Chapter one broaches this issue with a profile of Huston Smith, to whom mysticism is a kind of skylight through which all people in all eras can see the same transcendent reality. Chapter two highlights philosophers and theologians espousing what could be called a postmodern outlook; they contend that it is impossible to extract universal truths from the immense diversity of mystical experiences. Chapter three introduces the philosopher Ken Wilber, who rebuts the postmodernists with an "integral" worldview incorporating elements from both ancient mystical traditions and modern psychology.

These chapters lay the groundwork for the more scientifically oriented chapters that follow. Chapters four through eight profile scientists who have carried out empirical studies of mystical experiences, whether induced by meditation, prayer, epilepsy, electromagnetic stimulation of the temporal lobes, or psilocybin. These researchers include the aforementioned Andrew Newberg, Michael Persinger, James Austin, and Franz Vollenweider, as well as the British psychologist Susan Blackmore, who has scrutinized the paranormal propositions often advocated by mystical enthusiasts.

Some readers may be surprised — and dismayed — that I pay so much attention to psychedelics, or entheogens, as they are sometimes called. One reason is that these compounds give scientists a handhold on a slippery topic. "Psychedelic drugs are easier to study by the methods of modern science than most other means of inducing altered states of consciousness," the Harvard scholars Lester Grinspoon and James Bakalar stated in their book *Psychedelic Drugs Reconsidered,* "since they have known chemical structures and can be administered repeatedly under uniform experimental conditions."

Moreover, research on psychedelics, which largely vanished during the 1970s and 1980s, has recently undergone a renaissance, and it is now yielding some of the most provocative findings in the field of mystical research. My inquiries also convinced me that psychedelics — for good or ill — have played a surprisingly large role in shaping the landscape of modern spirituality; psychedelic epiphanies catalyzed the spiritual evolution of many mystics who now advocate nonpsychedelic practices and even disparage entheogens. Finally, psychedelicists such as the psychiatrist Stanislav Grof and the postmodern shaman Terence McKenna — the subjects of chapters nine and ten, respectively — have fashioned their hallucinatory visions into cosmologies too provocative to ignore.

The evolutionary biologist Edward O. Wilson has decreed that you cannot tread the path of spirituality and the path of reason; you must choose between them. One of my goals in writing this book was to put Wilson's dictum to the test. Thus, interviewing those with firsthand mystical experience, I put various questions to them to gauge how successfully they have integrated their mystical and rational perspectives: Do they adhere to the mystical doctrine that mind is more intrinsic to reality than matter? Do they believe in an afterlife? Have they intuited a divine intelligence or plan underlying the universe, a plan in which we humans play a central role? If so, can they explain why this plan involves so much seemingly gratuitous human suffering? In short, how do they reconcile their mystical beliefs with the dictates of science and common sense?

Heaven, hell, and visions

If I lay bare others' prejudices, it seems only fair that I do the same for myself. I have always been prone to eschatological obsession. As a child, I saw death as an unmitigated evil. After a classmate died when I was in first grade, I became preoccupied with death, and I could not understand why everyone wasn't equally preoccupied. My parents, siblings, friends, all were doomed, and yet they blithely went on with their lives as if they had all the time in the world. My horror of mortality was most acute in the most cheerful, chattering contexts — in a classroom or at a party. I wanted to scream out to the oblivious fools around me, "You're all going to die!"

My view of death is slightly more nuanced now. In 1986, just before my mother had an operation for brain cancer, I visited her in the hospital. Lying on her bed, she urged me not to worry about her. She had had a good life. She married a good man, and she got to see her five children grow up and thrive. When she told me that she had no fear of death, I believed her. She seemed serene, ready for whatever came. My mother survived the operation. When she died more than two painful years later, I saw it as a blessing. Now death per se does not trouble me so much as the manner in which it sometimes descends. Although some die peacefully after long, rich lives, others are wrenched away from life in such a brutal and untimely fashion that they leave behind a terrible wound. How can this apparent unfairness of existence be reconciled with our spiritual intuitions of a just, loving God or of a supernatural moral order?

Ordinarily, I would prefer to treat this problem as an intellectual puzzle, like the nature-nurture conundrum or the irreconcilability of quantum mechanics and general relativity. As I wrote this book, however, events seemed to conspire to remind me of fortune's terrible capriciousness. On September 11, 2001, my wife and I climbed a hill near our home and saw only smoke where once the World Trade Center had stood, fifty miles south of us. But that cataclysm was almost too vast, too singular, for me to fathom; other, more ordinary incidents had a deeper emotional impact. In the span of a year, several friends and acquaintances were diagnosed with cancer. One of my oldest friends died, leaving two young children behind. A mystical expert only slightly older than me and brimming with wit and vitality when I interviewed him was killed less than a year later by a malignant brain tumor.

Then there was the death of Lena, my family's familiar, on the day before I flew west to ingest the hallucinogenic brew ayahuasca. As hard as it must be for outsiders to understand, that little tragedy was a grievous reminder to me that happiness can be snatched from us at any time; the happier we are, the greater our potential heartbreak. When I ingested ayahuasca three days later, the hopes and fears that mysticism arouses in me came to a head. For these reasons, I decided to tell the story of that ayahuasca session in chapter eleven. Before the session, I hoped that it might give me an insight or epiphany or *something* that would provide consolation — not only for me but also for Suzie, who loved Lena dearly.

At the same time, memories of a nightmarish drug trip — the one to which I alluded in *The End of Science* — made me fear that ayahuasca might exacerbate my dread. The German psychologist Adolf Dittrich has compiled evidence that altered states — whether induced by drugs, meditation, hypnosis, sensory deprivation, or other means — fall into three broad categories, or "dimensions." Borrowing a phrase that Sigmund Freud used to describe mystical experiences, Dittrich calls the first dimension "oceanic boundlessness." This is the classic blissful, unitive experience reported by Richard Bucke and many other mystics. The mystic has sensations of self-transcendence, timelessness, and fearlessness, and an intuition that all the world's contradictions have been resolved.

Dittrich labels the second dimension "dread of ego dissolution." This is the classic "bad trip" in which your sense of self-dissolution is accompanied not by bliss but by negative emotions, from mild uneasiness to full-blown terror and paranoia. You think you are going insane, disintegrating, dying. Dittrich dubs the third dimension "visionary restructuralization"; it includes hallucinations ranging from abstract, kaleidoscopic images to elaborate dreamlike narratives. Dittrich likes to refer to these three dimensions as "heaven, hell, and visions."

The hallucinogen I ingested in 1981 propelled me into all three dimensions. Early on, I had fantastical, dreamlike visions teeming with animals, humans, and mythological figures. I seemed to be both observing these epic scenarios and playing all the parts in them. These images became more and more abstract and ethereal, until I became convinced that I was approaching absolute reality, the source of all things, God. Like Richard Bucke, I saw, I *knew*, that there is no death, not for me, not for anyone or anything; there is only life, forever and ever. Then the ground of being was yanked from under me. I saw, I *knew,* that life is ephemeral; death and nothingness are the only abiding certainties. We are in perpetual free fall, and there is no ground of being, no omnipotent God to catch us.

Hindus call truth that is perceived directly *shruti;* the Sanskrit term *smriti* refers to truth known only secondhand. Time transformed my 1981 experience from *shruti* into *smriti,* a memory of a memory of a memory. It resembles a faded photograph from a journey I can scarcely remember. It almost seems as though it happened to someone else. My years as a science writer also infused me with a skepticism so corrosive that it eroded my belief in all revelations, including my own.

I never forgot that trip, however, or stopped brooding over its implications. Which of our mystical visions should we believe? The heavenly, blissful ones or the hellish, diabolical ones? Are both somehow true, or are all such visions illusions, generated by overexcited neural circuits? Some mystically inclined philosophers — notably Huston Smith — have proposed an answer to questions like these. They contend that mystical experiences, in spite of their diversity and apparent contradictions, all point to the same universal truth about the nature of reality, a truth that is not frightening but comforting. This position is known as the perennial philosophy, and it is where we will begin this mystical inquest.

1

Huston Smith's Perennial Philosophy

In April 1999, I traveled to Albuquerque, New Mexico, to attend a meeting named, misleadingly, "Science and Consciousness." Held in a hotel modeled after a Mayan pyramid, the five-day conference was actually a New Age bazaar, serving up diverse products for boosting physical, mental, and spiritual energy. The five hundred or so attendees could jump-start their day with Vipassana meditation, tai chi, or a "dance of universal peace." Later they listened to speakers — most of whom hailed from alternative teaching centers such as the California Institute for Integral Studies and the Institute of Noetic Sciences — propound the benefits of Transcendental Meditation, shamanic drumming, and devotional prayer. In speech after speech, mystical experiences were invoked as a panacea that can heal both mind and body.

To the extent that lecturers invoked science, it was generally in the spirit of the New Age classic *The Tao of Physics*, in which Fritjof Capra proclaimed that modern physics has rediscovered mystical truths embodied in Hinduism and Buddhism.* That was the theme of a talk by Edgar Mitchell, an Apollo astronaut whose vision of the earth against the backdrop of space in 1971 convinced him that life was not just an accident but was divinely ordained. Mitchell went on to found the Institute of Noetic Sciences, which promotes science with a spiritual dimension.

---

*Capra has apparently backed away from the implications of *The Tao of Physics* (Shambhala Publications, Boston, 1975). In his essay "Science and Mysticism" (in *Paths Beyond Ego*, edited by Roger Walsh and Francis Vaughan, Jeremy Tarcher/Putnam, New York, 1993, pp. 189–190), Capra rejected the notion that modern science is "merely rediscovering ancient wisdom, known to the Eastern sages for thousands of years." There can be no "synthesis" between modern science and mysticism, he declared, because the two approaches are "entirely different" and at best "complementary."

Superstring theory and other esoteric advances in physics, Mitchell assured us in Albuquerque, are revealing that our reality is embedded in a much more expansive, higher-dimensional realm of pure energy — or pure spirit, as Mitchell preferred to think of it. The discovery of quantum nonlocality — the ability of particles to exert subtle influences on each other instantaneously across vast distances — is confirming the ancient mystical teaching that all things are profoundly interconnected. Quantum nonlocality might also explain extrasensory perception, Mitchell added, as well as the miraculous healing that results from prayer and other spiritual practices.

Part of me bridled at these efforts to transform physics into a feel-good religion. I tended to agree with the physicist and Nobel laureate Steven Weinberg that "as we have discovered more and more fundamental physical principles they seem to have less and less to do with us." At the same time, I sympathized with the sentiments on display in Albuquerque. Speakers and audience members alike exuded an almost palpable yearning for the transcendent, the marvelous, the mystical.

One lecturer, after giving a talk on the links between "quantum biology" and Eastern mysticism, led us in guided meditation. He turned on a tape of tinkling music, set a turquoise-colored vase on a table, and instructed us to "resonate with the vase's emptiness." Afterward, an audience member enthused with a Texas twang that he felt the vase "turn into particles and disappear." "I had that experience too," the lecturer replied, adding that at these moments we are "experiencing God."

Stepping outside the hotel for some fresh air, I found myself in a throng of smokers. I overheard a heavyset, blond-bunned woman telling a friend about an "amazing" experience she had had earlier that day. While relaxing with her eyes closed in the hotel's hot tub, the blond woman recalled, all her boundaries dissolved, and she felt as though she were a fetus floating in a cosmic womb. The friend looked at her with a mixture of skepticism and envy, as did I.

In one of the more scientifically rigorous lectures, the ethnopharmacologist Dennis McKenna, the younger brother of the psychedelic provocateur Terence, discussed the mind-bending properties of dimethyltryptamine, or DMT. As I took a seat before McKenna's talk, a petite, white-haired woman in a sensible pantsuit asked me if I knew anything about this "DMT substance." I explained that DMT is an extremely powerful hallucinogen, which when smoked or injected in its pure form delivers the equivalent of a ten-hour LSD trip compressed into less than half an hour. I added that a DMT-laced beverage called ayahuasca serves as a sacrament for churches in South America and elsewhere. The grandmotherly woman nodded earnestly and asked, "Is the speaker going to tell us how to obtain it?"

I came to this meeting in part to steep myself in mysticism, New Age style. But a more important goal was meeting Huston Smith, an eighty-year-old religious scholar who had recently retired from the University of California at Berkeley. Smith is a walking, talking embodiment of the perennial philosophy. The seventeenth-century German philosopher Gottfried Leibniz coined the phrase *philosophia perennis,* and it was popularized in the twentieth century by the British author Aldous Huxley (who wrote a book titled *The Perennial Philosophy*) and others.

The perennial philosophy holds that the world's great spiritual traditions, in spite of their obvious differences, express the same fundamental truth about the nature of reality, a truth that can be directly apprehended during a mystical experience. Implicit in the perennial philosophy is the notion that mystical perceptions transcend time, place, culture, and individual identity. Just as a farmer in first-century China and a Web site designer in twenty-first-century New York City see the same moon when they look skyward, so will they glimpse the same truth in the depths of a mystical vision.

Raised by Methodist missionaries in China, Smith earned a doctorate in philosophy from the University of Chicago in 1945 and went on to become one of the twentieth century's foremost philosophers of religion. His 1958 book *The World's Religions* (whose original, pre-feminism title was *The Religions of Man*) has sold more than two and a half million copies and remains a definitive text of comparative religion. Reading the book, I understood why it has fared so well. Smith's writing style was unusually vivid for a serious religious scholar. He was passionate, witty, and scrupulously honest about his intentions.

Smith admitted in his introduction that he would not dwell on the many sins committed in the name of religion, including "human sacrifice and scapegoating, fanaticism and persecution, the Christian Crusades and the holy wars of Islam." His goal was not to criticize religions but to promote understanding of them by showing them in the best possible light. The book was animated by affection not only for Smith's personal faith,

Christianity, but for all "wisdom traditions." Far from downplaying religions' differences, Smith celebrated them. He likened religions to "a stained glass window whose sections divide the light of the world into different colors."

At the end of *The World's Religions,* Smith addressed the problem that the perennial philosophy seeks to solve. Given the enormous diversity of religious beliefs, how should we choose among them? Smith's response was that of a grandfather confronted with grandchildren demanding to know whom he loves most: I love you all! he replied, so sincerely that everyone can go away happy. Smith justified his response by arguing that all religions agree on three fundamental tenets: First, reality is more unified than it appears. Second, reality is better than it ordinarily seems to us. Third, reality is more mysterious than it looks. These are the central insights of the perennial philosophy, which mystics know firsthand.

An entheogenic epiphany

When I started my research for this book, Smith's name kept coming up in my readings and interviews; he was described as an authority on mysticism in both the scholarly and spiritual sense. Smith has not been content to know other wisdom traditions only from the outside. In the mid-1950s he learned yoga and meditation from a Hindu swami. From 1958 to 1973 he practiced Zen meditation, and for the next fifteen years he plunged into Sufism, the mystical tradition of Islam. In the late 1980s he befriended members of the Native American Church and participated in their ceremonies. After one of his daughters married a Jew and converted to Judaism, Smith steeped himself in that faith.

Smith's writings so impressed the journalist Bill Moyers that he produced a five-part series of interviews with Smith, broadcast in 1996 as *The Wisdom of Faith with Huston Smith.* The more I heard about Smith, the more he seemed the ideal person to help me get my bearings as I ventured into the deeps of mysticism. It was thus with great anticipation that I arrived at the "Science and Consciousness" conference in Albuquerque, just in time to hear Smith's lecture.

To my immense disappointment, I was turned away at the door of the ballroom; there was not even standing room. I lurked by the entrance like a groupie waiting for Smith to emerge. When he finally appeared, he was mobbed by admirers. A tall man, he loomed above most of those around him. His outfit was vaguely Asian, vaguely clerical: white pants, white collarless shirt, and a woven vest. Smith had snowy hair, a close-cropped beard, and a bearing both gentle and regal. He wore a hearing aid in each ear, and he had a slight stoop, perhaps the result of his habit of leaning intently toward anyone speaking to him.

Smith listened and responded patiently to each person as I stood, increasingly exasperated, outside the throng. During a lull in the discussion I thrust my hand at Smith, introduced myself, and asked if he had time for an interview. To my delight, he suggested that we meet for dinner in his room. As we sat down that evening before our room-service meals, Smith paused for grace. "As they say in the Zen monastery, *Itadakimasu,* which is the world's shortest blessing and therefore my favorite. All it means is 'I eat.'" Cutting into his lamb, Smith recalled that his old friend Joseph Campbell, when asked what his "yoga" was, had mentioned three things: rare roast beef, good Irish whiskey, and forty-four laps in the pool every day.

Smith spoke in slow, measured tones. His voice, in spite of an occasional quaver, was strong and resonant, and he laughed frequently. He seemed to derive sensual pleasure from words — particularly exotic ones, such as *congeries* — savoring them syllable by syllable. Smith exuded intelligence, humor, and generosity of spirit — all the traits that one looks for in a sage — but he could be sharp-tongued, too. He even had a few harsh words for the world's religions, or "wisdom traditions."

"I'm the first to insist that not everything about them is wise," he said. The wisdom traditions have in the past legitimized caste divisions, slavery, and the subjugation of women; the religious leadership in the United States, "particularly in the dominant religion, Christianity, is third rate." Many young people turn away from religion, Smith lamented, because it is presented to them in the worst possible way. "What came through to them were two things: dogmatism, 'We've got the truth, everybody else is going to hell'; and moralism, 'Don't do this, that, and *especially* not the other thing.'" Smith's missionary parents had imparted very different religious precepts to him: "First, we're in good hands. And second, in gratitude for that, it would be well if we bore one another's burden." After a lifetime of studying many religions, Smith had found no better formula.

Religious institutions are in some sense a necessary evil, Smith said. "If you do not institutionalize your spirituality, it gets no traction on history. So if Jesus had *not* been followed by Saint Paul, who established a church, the Sermon on the Mount would have evaporated within a generation. So I think if we are serious about the spiritual life, we have to undertake the burdens of institutionalization — and they are heavy."

Religions, Smith asserted, serve as our major repository of the perennial philosophy. They all agree that beyond this mundane, material world is a transcendent realm; although we can in rare moments glimpse this reality, it is as distant from our descriptions of it as a finger pointing skyward is distant from the moon. "The fact that they came to that view independently," Smith continued, "in India, China, and East Asia, and the Abrahamic traditions, lends a certain, in my mind, *a priori* or *prima facie* sense that it might be right — at least in fitting the human makeup."

When I asked Smith about his own mystical experiences, he chose his words with extra care, for reasons that soon became apparent. "I'm pretty flat-footed as a mystic," he said. "I'm not good at moving into a distinctly different state of consciousness." Smith had been meditating for almost half a century, and he still meditated almost every day. But his meditative experiences were "pretty ordinary, garden variety."

His most important mystical experiences, Smith told me, were "entheogenic." The word literally means "God-containing" and was popularized in the 1980s by Gordon Wasson, a banker turned mushroom expert, as a term for plants or chemicals that induce spiritual experiences. Smith prefers the term *entheogenic* to *hallucinogenic*, which he considers derogatory and inaccurate, and to *psychedelic*, which is too closely associated with the 1960s and recreational drug use.*

Smith has been cautious about divulging his entheogenic encounters. Fearing misunderstanding, he did not mention them in his broadcast conversations with Bill Moyers. Smith emphasized to me that mind-expanding

---

*Although Wasson is often credited with coining *entheogen*, it was actually introduced by the Greek scholar Carl Ruck and his coauthors in "Entheogens" (*Journal of Psychedelic Drugs*, vol. 11, nos. 1–2, 1979, pp. 145–146). As for *psychedelic*, Humphrey Osmond, a psychiatrist and pioneer of research on LSD, coined the term in 1956, as he recalled in his book *Predicting the Past* (Macmillan, New York, 1980, pp. 81–82). Dissatisfied with *psychotomimetic* and *hallucinogen*, Osmond asked his "dear friend" Aldous Huxley if he could suggest an alternative for drugs such as mescaline and LSD. Huxley proposed *phanerothyme*, from the Greek words for "soul-revealing." Huxley wrote: "To make this trivial world sublime / Take half a gramme of phanerothyme." Humphrey countered with *psychedelic*, from the Greek roots for "mind-revealing," and with another ditty: "To fall in hell or soar angelic / Just take a pinch of psychedelic." Osmond credited "naughty" Timothy Leary with popularizing the term *psyche-*

substances can be dangerous and should be used with great care and reverence. Entheogens alone do not constitute a spiritual path. They can at best yield glimpses of the transcendent realm; they cannot deliver you to that realm permanently. They may also encourage an unhealthy obsession with altered states, Smith said, when the goal of spirituality should be the transformation of one's whole life.

Smith took entheogens in the early 1960s, when he was teaching at the Massachusetts Institute of Technology, but he stopped after they were outlawed in 1966. He has ingested mind-expanding compounds during only one period since then, in the early 1990s, when he participated in four legal peyote ceremonies of the Native American Church. Smith has no desire to take entheogens again. "I'd had the message," he said, "and I felt no wish to go back."

But Smith clearly cherishes his entheogenic experiences. His first took place on New Year's Day, 1961, in Newton, Massachusetts, at the home of Timothy Leary, then a Harvard psychology professor just beginning to investigate psychedelics. Leary gave Smith two capsules of mescaline. A few hours later Smith felt he was witnessing the reality described in the ancient Hindu Vedas and other mystical texts. He was seeing through the mundane reality around him to the ground of being, the clear light of the void underlying all things.

"From the soles of my feet on," he recalled, "I found myself saying, 'Yes! Yes!'" The experience was not entirely pleasant; Smith once described it as "strange, weird, uncanny, significant, and terrifying beyond belief." Toward the end of the trip, Smith approached Leary and said, "I hope you know what you're playing around with here . . . There *is* such a thing as people being frightened to death."

Mysticism versus scientism

One distinguishing characteristic of Smith's spiritual philosophy is his view of enlightenment. He defines enlightenment as "a quality of life" rather than a state of mind; it entails not altered states but altered *traits*. To

---

delic. In *DMT: The Spirit Molecule* (Park Street Press, Rochester, Vermont, 2001), the psychiatrist and psychedelic researcher Rick Strassman listed alternatives to *psychedelic* coined over the years: *cultogen, mysticomimetic, oneirogen, phantasicant, psychodysleptic, psychotogen, psychotoxin, schizotoxin* (pp. 30, 348).

Smith, mature mystical knowledge must manifest itself throughout one's life. "If you think you are advancing toward unity with God or the absolute," he said, "and are not growing in love and charity toward your fellow person, you're just deluding yourself." Smith would be "profoundly suspicious" of anyone who claimed to be enlightened but did not exhibit these basic human virtues.

In fact, Smith sees enlightenment as "an ideal" that can be approached but never fully attained by any mortal. After he voiced this opinion in his televised dialogue with Bill Moyers, Smith received letters proclaiming the enlightenment of various spiritual masters, such as the American guru Da Free John and the Indian swami Sai Baba. Smith remained unmoved. "Even Christ said, 'Why callest thou me good?'" he explained to me. "Does anybody presume to top that?" Smith shook his head. "Just in principle, I don't think it's possible," he said. "The mortal coils are too tight."

The irony is that Smith himself is viewed by some as having reached the highest echelons of spirituality. The transpersonal psychologist Frances Vaughan has written that Smith epitomizes "mature spirituality," and the mythologist Jean Houston described him to me as "a person of very deep wisdom and compassion." I cannot dispute these assessments. When Smith turned his beaming, inquisitive face toward me in Albuquerque, I had to squelch a compulsion to bow. Reverence is not an emotion that interviewees ordinarily evoke in me.

That said, I have difficulty accepting certain aspects of his philosophy. First, I find Smith too forgiving of religions' faults and contradictions. His affection for religions is of course one of his most appealing traits. He does not merely tolerate the diverse faiths of the world; he embraces them in all their multifarious glory. But different religions say different things about God, life after death, and ultimate reality. These doctrines can't all be right, can they? Smith suggests that they can, somehow. He once wrote that different religions "'flesh out' God's nature by seeing it from different angles. They supplement our view without compromising the fact that each angle is, in its own right, adequate, containing (in traditional locution) 'truth sufficient unto salvation.'"

All these contradictions and distinctions, Smith asserts, dissolve within the supreme mystical vision, which reveals a reality that transcends all worldly categories of experience and description. This vision is ineffable, or apophatic, a theological term that means devoid of any specifiable content. The apophatic experience "goes deeper noetically and philosophically into the human apprehension of the deepest reality," Smith told me.

His conclusion is based in part on his first entheogenic trip in 1961, in which he glimpsed the pure light of the void in Timothy Leary's living room. This sensation seemed to bring Smith closer to absolute truth than visions yielding more specific insights. Smith's belief that apophatic experiences yield the deepest truth also relied on philosophical reasoning. A vision that attributes particular properties to ultimate reality "is as much about what it excludes as about what it is," Smith explained. "But the infinite can exclude nothing. And therefore when these lines of demarcation dissolve, I have a feeling I am closer to how it is."

One can ask whether Smith's 1961 vision was truly apophatic and ineffable. After all, he did describe it, and it yielded quite specific insights. Smith has written that the experience confirmed a metaphysical theory known as emanationism, which posits that the clear light of the void "fractures into multiple forms and declines in intensity as it devolves through descending levels of reality." But that issue aside, how can we infer anything about ultimate reality from a truly apophatic experience? More specifically, how can we know that reality is better than it seems, or that we are "in good hands," as Smith has put it?

My major misgiving about Smith concerns his attitude toward science. I see science, in spite of its limitations, as a polestar that we desperately need for guidance when we venture onto the perilous waters of mysticism. Smith once had a positive view of science. As a child, he saw science's power to relieve human suffering when he watched his missionary father administer smallpox vaccines to his Chinese wards. In graduate school, Smith still believed that theology and philosophy had to take their leads from science, which is our best guide to truth.

But over the course of his career Smith has become increasingly ambivalent about science. In his 1976 book *Forgotten Truth,* he contended that modern physics and cosmology are corroborating the perennial philosophy by revealing the cosmos to be larger, more mysterious, and more unified than our ancestors ever imagined. Having appealed to the authority of science, however, Smith went on to denounce its excesses. He claimed that his quarrel was not really with science per se but with *scientism,* the ideological insistence that science is not only our best but also our sole means of understanding reality. Yet Smith's critique occasionally devolved into condemnation of science's basic purpose. "The Western hunt for knowledge, analytic and objective to its core, has violence built into it," he stated. "For to know analytically is to reduce the object of knowledge, however vital, however complex, to precisely this: an object."

When I asked Smith why he had become so annoyed by science, he smiled ruefully at my question's implication: "We need a good therapist to understand Huston!" He traced his annoyance back to his fifteen-year stint at that temple of scientism, MIT. "I never had any trouble with the scientists, or the students, all of whom were science majors. But my fellow philosophers drove me up the wall." The philosophers worshiped science as the ultimate mode of knowledge, and they disparaged scholars like Smith who value religious propositions. "I knew I could run out the clock just trying to get them to grant that I, with my different point of view, was even doing philosophy."

Science has made enormous contributions to human life; Smith knows that. He is nonetheless appalled that so many of us now view scientific truth as the only truth that matters. This "scientistic" ideology, he said, has "created a huge store of unnecessary human pain" by denigrating all the beliefs that have traditionally given meaning and purpose to our lives.

Although he worried that his complaints about science might sound whiny and mean-spirited, Smith felt compelled to deliver his message as forcefully as possible in a new book he was writing, called *Why Religion Matters.* "Let's put it this way," he said to me, his voice rising. "What about the prophets of Israel? When they confronted the misdoings of kings, would you expect them to have a modulated voice? When they say, 'Thus sayeth the Lord! Where the dogs licked up the blood of Naboth, there they will lick up your blood!' — you tell me, what's the appropriate stance?" Smith's voice trembled and his eyes blazed; he looked as fierce as Elijah, the Old Testament prophet he was quoting.

It isn't clear to me, I replied, just how much scientism has gripped modern society. Even if traditional religions are declining in influence, spirituality keeps bubbling up throughout our culture; this conference in Albuquerque where we were meeting provided abundant evidence of that. I added that I know few people, including scientists, who have a truly materialistic, atheistic view of the world. "If that is true, then I'm fighting a straw man," Smith said, so quietly that I regretted my words. "But I'm going to say it anyway. And if it's an irate, peevish book, okay." He shrugged, cheerful again. "That's one little register into the ongoing discourse."

The Good Friday experiment

Smith invokes science when it suits his purposes. For example, to bolster his claim that entheogens can induce genuine mystical experiences, he cites a study called the Good Friday experiment, in which he participated. Sometimes called the Miracle of Marsh Chapel, it took place in 1962, a time of intense debate over psychedelics' spiritual potential.

Aldous Huxley had made the case for psychedelic mysticism in dramatic fashion in his 1954 book *The Doors of Perception*. It described a day that Huxley spent intoxicated on mescaline, the active ingredient of peyote. Gazing at a vase of flowers in his California home, Huxley saw "what Adam had seen on the morning of his creation — the miracle, moment by moment, of naked existence." Huxley grasped mystical truths that he had previously glimpsed only from afar, through the narrow aperture of his intellect. He realized what the thirteenth-century German mystic Meister Eckhart meant when he talked about *Istigkeit*, or Is-ness, and what the Hindus intended by the phrase *Sat Chit Ananda*, or Being-Awareness-Bliss. Huxley wrote:

> I am not so foolish as to equate what happens under the influence of mescaline or of any other drug, prepared or in the future preparable, with the realization of the end and ultimate purpose of human life: Enlightenment, the Beatific Vision. All I am suggesting is that the mescaline experience is what Catholic theologians call a "gratuitous grace," not necessary to salvation but potentially helpful and to be accepted thankfully, if made available.

The experience shifted Huxley's attitude toward mind-altering drugs. In his pre-mescaline novel *Brave New World*, a drug that Huxley slyly called soma — after a divine potion mentioned in ancient Hindu texts — served as a soporific with which the state pacifies the masses. In Huxley's post-mescaline novel *Island*, inhabitants of a utopian nation called Pala took a mushroom-based compound not to escape reality but to see it in all its sublime glory. When Huxley called this substance moksha, the Sanskrit word for liberation, he intended no irony.

At least one prominent scholar strenuously objected to Huxley's suggestion that mescaline can trigger a genuine mystical experience. The Catholic theologian R. C. Zaehner of the University of Oxford contended that drugs at best enhance our appreciation of and identification with nature; they cannot bestow what Zaehner considered to be the supreme mystical experience, in which the mystic apprehends nature's transcendent creator, God. In his zeal to rebut Huxley, Zaehner consumed mescaline himself; the drug apparently gave him an acute case of the giggles.*

The Good Friday experiment was intended to bring empirical evidence rather than just anecdotal reports to bear on the debate over psychedelic drugs and mysticism. The experiment was conceived by Walter Pahnke, a psychiatrist enrolled in a Harvard doctoral program on religion and society. His supervisor was Timothy Leary. Pahnke selected twenty Protestant divinity students, all white males, to serve as subjects.

---

*During Zaehner's encounter with mescaline, which took place on December 3, 1955, at Oxford, he was accompanied by several of his fellow dons, whom he called "investigators." His book *Mysticism Sacred and Profane* (Oxford University Press, New York, 1961, pp. 217–218) presented transcripts of some of their tape-recorded conversations. At one point the investigators showed Zaehner Gentile da Fabriano's painting *Adoration of the Magi,* which shows the three wise men surrounding the infant Jesus. Zaehner became convinced that one of the Magi was about to bite the foot of Jesus, who was trying to push the old man away. Zaehner burst into laughter, and the following exchange ensued:

"What do you find so funny, Professor Zaehner?" a colleague asked.

Zaehner (ecstatically): "Nothing."

When his laughter subsided, Zaehner said, "You all look so serious."

Investigator: "We can't enjoy it to the same extent you can."

Zaehner (convulsed): "No, I suppose you can't." (Further gusts of uncontrollable laughter.)

A bit later, Zaehner said, "I wish everyone wasn't being so unfriendly."

Investigator: "Do we seem unfriendly?"

Zaehner (loudly and emphatically): "Frightfully." (Prolonged and uproarious laughter.)

"In Huxley's terminology," Zaehner reflected later (p. 226), "'self-transcendence' of a sort did take place, but transcendence into a world of farcical meaninglessness. All things were one in the sense that they were all, at the height of my manic state, equally funny." Zaehner denigrated the experience as "trivial" and "anti-religious." The Catholic theologian was either unfamiliar with or dismissive of the notion of the laughing Buddha, whose liberation from this vale of tears manifests itself as belly-jiggling mirth. "Good and evil reconciled in a laugh!" William James scribbled down just after inhaling nitrous oxide (as quoted in "The Nitrous Oxide Philosopher," by Dmitri Tymoczko, *The Atlantic Monthly,* May 1996, p. 93). James once ingested peyote, on June 8, 1896, but he experienced only nausea. In a letter to his brother he wrote, "I took one bud 3 days ago, vomited and spattered for 24 hours . . . I will take the visions on trust" ("Visions of the Night," by Daniel Perrine, *The Heffter Review,* vol. 2, 2001).

Ten students received a capsule of psilocybin, the active ingredient of psilocybin mushrooms, whose effects last from four to six hours. The other ten students ingested nicotinic acid, a so-called active placebo, which induces face-flushing and other minor physiological effects but does not affect cognition. The experiment was double blind; to eliminate bias, neither the subjects nor the researchers knew in advance who was receiving the psilocybin.

The twenty divinity students were accompanied by ten "group leaders" — including Huston Smith and other faculty members — who supervised the proceedings. Half of the group leaders received capsules of psilocybin and half received nicotinic acid, also blindly. For the most part, everyone stayed in a room in the basement of Marsh Chapel, on the campus of Boston University. Speakers piped in the sounds of a Good Friday service held in the main chapel above them. Immediately after the session, the subjects rated their experiences in terms of various mystical qualities, such as a sense of unity, sacredness, ineffability, and transcendence of time and space. Six months later, the subjects were given the same questionnaire, and they were asked about the session's aftereffects.

The findings reported by Pahnke were dramatic. The ten psilocybin subjects all ranked their experiences much higher in mystical qualities than members of the control group did. Six months later, the psilocybin group reported persistent beneficial effects on their attitude and behavior; the experience had deepened their religious faith, made them more loving and empathetic toward others, and sharpened their appreciation of life's wondrousness. The experiment was widely hailed as proof that psychedelic drugs can generate life-enhancing mystical experiences.

One participant who remembers the Good Friday experiment fondly is Huston Smith, who was one of the group leaders who received psilocybin. At the climax of the Good Friday service, a soprano sang a hymn that ended with this stanza:

My times are in Thy hands, I'll always trust in Thee;
And after death at Thy right hand I shall forever be.

The soprano's "angelic" voice catapulted Smith into what Hindus call a *bhaktic* experience. That is "a love relationship with the divine," he explained to me, "with God imaged in personal terms." Although previously Smith *believed* that "God is love and that none of love's nuances could be absent from His infinite nature," he was now feeling that divine love firsthand. The afterglow persisted for months. "I really was a better person," Smith said of this period. He had a vivid sense that life "is a miracle, every moment of it, and that the only appropriate way to respond to the gift we have been given is to be mindful of that gift at every moment, and to be caring toward everyone we meet."

But Smith also recalled some less exalted aspects of the Miracle at Marsh Chapel. While he and others who had received psilocybin were lost in mystical reveries, their sober comrades "felt left out and were not above acting out their resentment in derisive laughter and incredulous hoots over the way the rest of us were behaving."

As the psilocybin began working its magic on Smith, he caught the eye of a colleague who seemed to be similarly transported. Smith said merely, "It's true, isn't it?," assuming his colleague would intuit that he meant "the religious outlook, God and all that follows." Only after the experiment did Smith learn that his colleague had received the placebo pill, not the psilocybin, and had been baffled by Smith's remark. "So I was dead wrong in inferring from our eye contact that our minds were in sync."

A recent review of the Good Friday experiment has raised more serious questions about it. The review was carried out by Rick Doblin, a graduate of Harvard University's School of Government and an advocate for the reform of laws banning mind-expanding drugs. Beginning in the late 1980s, Doblin tracked down and interviewed twenty-three of the thirty subjects and group leaders who participated in the experiment.

Doblin's investigation confirmed many of Pahnke's original findings. Most of those given psilocybin still felt that it "had significantly affected [their lives] in a positive way," Doblin summarized in *The Journal of Transpersonal Psychology.* The experience "helped them to resolve career decisions, recognize the arbitrariness of ego boundaries, increase their depth of faith . . . and heighten their sense of joy and beauty." But only two psilocybin subjects described the experiment as totally positive. Most recalled "moments in which they feared they were either going crazy, dying, or were too weak for the ordeal they were experiencing," Doblin wrote.

Pahnke, who died in 1971, had downplayed these negative reactions in his published reports and omitted any mention of a subject known as L.R. Initially overwhelmed with fear and distrust of those around him, L.R. then became convinced that God had chosen him to announce the imminence of an age of peace. He burst out of Marsh Chapel and fled down the busy street outside to spread the word. Three group leaders, including Huston Smith, ran after L.R. and succeeded in bringing him back to the chapel basement. He remained so agitated that Pahnke injected him with the anti-psychotic drug thorazine. Long after the session, L.R. continued to have anxiety attacks.

In other words, the Good Friday experiment showed that an entheogen can induce mystical experiences with lasting positive effects in subjects who are already religiously inclined and are in a safe, supervised, religious setting. But even under these ideal conditions, an entheogen can also trigger extreme anxiety and delusional psychosis (although L.R.'s delusion, that he was God's specially chosen messenger, has been shared by many religious leaders).

Rick Doblin could find "no justification" for Pahnke's failure to mention L.R.'s reaction and other negative aspects of the Good Friday experiment. Doblin even suggested that Pahnke's whitewashed description of the Miracle of Marsh Chapel, which was widely reported in the media, contributed to the careless consumption of psychedelics in the 1960s and hence to the anti-drug backlash.

The Good Friday experiment could be said to symbolize the promise and pitfalls of mystical experiences in general. They can be blissful and life-enhancing, but they can also trigger paranoia, narcissistic delusions, and other forms of madness. A Talmudic legend about four learned rabbis who are allowed to visit paradise makes this point: One rabbi dies outright, one goes mad, and one becomes a heretic. A single rabbi leaves heaven with a blissful, peaceful heart, his faith confirmed.

Even after the most positive mystical experiences, we are left with the task of reconciling those visions with what our reason tells us about the world. For example, if there is a God, and if He loves us, as Huston Smith felt so strongly during the Good Friday experiment, why did He create a world with so much evil and injustice and suffering? When I put this age-old theological conundrum to Smith, his expression darkened. "I think there is an answer," he replied. "There is no way to prove it. It's pretty subtle."

He warned that his theodicy — or attempt to explain why a benevolent God allows evil — might not make any more sense than the abstruse theories of modern physics. "One of my frustrations is that everybody recognizes that it takes about five years to get your head into thinking about relativity theory and quantum mechanics. It's just such a different world. But everybody assumes that the religious world is accessible to everyone. Now, I profoundly disagree with that."

William James, after inhaling nitrous oxide, envisioned good and evil as two aspects of the same whole — "*but* with the good being the larger genus of which evil was the smaller species," Smith said. "I think that points in the direction of the answer." We are so immersed in evil and suffering, he said, that we give them great weight; in higher states of awareness, we can see that evil is just one component of a reality that is fundamentally beautiful and good.

If a child drops her ice cream cone on the ground, "it's the end of the world. Tears! Now, the mother can feel the pain of the little girl, but she knows that's not the end of the world. It is encompassed as a learning experience." Smith paused. "Ultimately . . ." He paused again, frowning. "This sounds so bad that I almost feel I shouldn't say it, and only because you're serious I will." Ultimately, he said, "even Auschwitz can be enveloped into a meaningful framework. But I don't say that very often. Because if one has to choose between really going into the experience of evil or just sort of skimming over the surface in favor of some grandiose picture which will resolve it, it's better to go into the evil. Otherwise, we enter into psychological denial."

So evil is in some sense necessary? "Yeah," Smith replied flatly. "Because it is part of finitude. If there were only the divine perfection, there would be no evil. But then the divine would not be infinite, and that's contradictory, too. So the divine *has* to include *every* level of being. And once you get a hairsbreadth away from absolute perfection, then evil enters."

But I still can't understand, I persisted, why this infinitely powerful God allows children and other innocents to know nothing but pain in their lives. That's the sticking point for me. "Rightly so," Smith murmured, nodding. He told me about a friend who recently learned from her doctor that an eye disease would soon blind her. After leaving the doctor's office, she visited a minister, whom Smith also knows, to ask for spiritual guidance. When she began weeping, the minister said nothing; he just took her hand and wept with her. "That's the only appropriate response if one is really in the face of that kind of thing," Smith said somberly. "Anything else is glib."

As I mulled over Smith's words later, this story kept coming back to me. He seemed to be acknowledging that there is, finally, no belief, doctrine, or dogma that can justify God's ways. Our mutual compassion is our only real consolation.

God's multiple-personality disorder

Mystical revelations are often said to be so self-validating that someone lucky enough to have one never doubts its veracity thereafter. That was not the case for William James. Under the influence of nitrous oxide, he did indeed briefly glimpse a realm in which all the world's contradictions were resolved and evil vanished. But James had difficulty reconciling this vision with the world in which he lived.

In *The Varieties of Religious Experience,* James raised doubts about Huston Smith's faith that we are all "in good hands"; after all, "melancholic" visionaries see the world as meaningless or malignant. James recounted some "diabolical" visions in *Varieties,* including one attributed to an anonymous French physician. The physician was alone in his dressing room when he suddenly recalled an epileptic patient he had once encountered in an insane asylum. The patient was a "black-haired youth with greenish skin, entirely idiotic . . . moving nothing but his black eyes and looking absolutely non-human. *That shape am I,* I felt, potentially," the physician thought.

The physician was describing a unitive experience of sorts, but his identification with the epileptic youth — his realization of what Hindus call Thou Art That — triggered not mystical bliss but "horrible dread." All that separated him from the insane boy, the physician recognized, was luck. He was left with an excruciating awareness of "the pit of insecurity beneath the surface of life."

Only after *Varieties* was published did James admit to a translator that the vision of the epileptic youth was actually his. James struggled throughout his life with depression and panic attacks, during which he recoiled from the randomness, injustice, and meaninglessness of existence. He sought relief in Christian Science and various "mind cures" that extolled positive thinking as the route to mental health, but nothing worked.

James was obsessed with what theologians sometimes call natural evil, suffering caused not by human malice but by what could be called acts of God, including congenital brain defects, earthquakes, tempests, and all the other random vicissitudes of life. Theologians often distinguish natural evil from the crimes that we humans deliberately perpetrate against each other, such as the Holocaust and slavery. But all evil is arguably natural evil, since God — if He exists — instilled the capacity for evil in us.

One of the most extreme reactions to the problem of natural evil was gnosticism. Gnostics were a diverse group of quasi-Christian mystics who flourished in the first few centuries after Christ's death but were eventually purged from the nascent church. Gnostics proposed a dizzying array of doctrines, and their behavior ranged from asceticism to licentiousness. But the essential gnostic theodicy is that something is drastically wrong with our world. The world was created not by the perfect, all-powerful, loving God postulated by conventional Christianity — and by perennialists such as Huston Smith — but by a flawed or even demonic deity. Gnostics called this deity the Demiurge, a name originally coined by Plato to describe the creator of the base material world (as opposed to the sublime world of forms).

The theologian Bernard McGinn, an authority on Christian mysticism, says that gnosticism reflects a "cosmic paranoia" or "demonization of the cosmos." Mystics, Aldous Huxley once said, see that in spite of all the pain and injustice in the world, everything is "All Right." Gnostics' mystical visions propelled them to the opposite conclusion: Everything is All Wrong.

The entheogenic journey I took in 1981 made me sympathetic toward gnosticism. The trip occurred in early summer, just after I had finished my junior year of college. I had left my apartment in New York City to visit friends in suburban Connecticut. One of these friends, whom I'll call Stan, was a psychedelic aficionado with an unusual connection: a chemist who investigated psychotropic drugs for a defense contractor in Research Triangle Park, North Carolina. The chemist had recently given Stan a thimble's worth of beige powder. It was supposedly a kind of supercharged LSD, although not as potent by weight.*

---

*After hearing me describe this drug's effects, John Halpern, a psychiatrist at Harvard University and an authority on psychedelics, guessed it was 3-quinuclidin-3-yl benzylate, otherwise known as BZ, or an analog thereof. BZ is a powerful hallucinogen that was developed as a chemical "incapacitant" by the U.S. Army in the 1950s. Although BZ was apparently never deployed, the Army stockpiled canisters of the drug until at least the early 1970s, when President Richard Nixon ordered the stockpiles destroyed. BZ is said to incapacitate subjects for up to eighty hours.

Naturally, I was curious, as was Stan, who had not sampled the stuff yet. One morning we each ingested about a matchhead's worth, a dose Stan thought the chemist had recommended. Within a half hour, I knew that this was by far the most powerful drug I had ever taken. I felt as though a volcano were erupting within me. Sitting on a lawn, barely holding myself upright, I told Stan that I feared I had taken an overdose. Stan, who for some reason was less affected by the compound, tried to calm me down. Everything would be fine, he said; I should just relax and go with the experience. As Stan murmured reassuringly, his eyeballs exploded from their sockets, trailed by crimson streamers.

That was my last contact with external reality for almost twenty-four hours. Stan and a couple of friends whose help he enlisted told me later that during this period I was completely unresponsive to them, although they could with some difficulty move me around. For the most part, I lay or sat quietly, staring into space. Occasionally I flailed about, raving, grunting, or emitting other uncategorizable sounds. For a long time I hissed like a five-year-old boy pretending to be a jet fighter: "*Fffffffffff!*" My expressions tended toward extremes: beatific, enraged, terrified, lewd. I sometimes furiously clawed holes in the lawn. My eyes remained wide open, the pupils dilated to the rim. My companions said I never seemed to blink, even when particles of dirt from my excavations were visible on my eyeballs.

Subjectively, I was immersed in a visionary phantasmagoria. I became an amoeba, an antelope, a lion devouring the antelope, an apeman squatting on the veldt, an Egyptian queen, Adam and Eve, an old man and woman on a porch watching an eternal sunset. At some point, I attained a kind of lucidity, like a dreamer who realizes he's dreaming. With a surge of power and exaltation, I realized that this is *my creation, my cosmos,* and I can do anything I like with it. I decided to pursue pleasure, pure pleasure, as far as it would take me. I became a bliss-seeking missile accelerating through an obsidian ether, shedding incandescent sparks, and the faster I flew, the brighter the sparks burned, the more exquisite was my rapture. This was probably when I was making the *fffffff* noise.

After what seemed eons of superluminal ecstasy, I decided that I wanted not pleasure but knowledge. I wanted to know *why.* I traveled backward through time, observing the births and lives and deaths of all the creatures who had ever lived, human and nonhuman. I ventured into the future, too, watching as the earth and then the entire cosmos was transformed into a vast grid of glowing circuitry, a computer dedicated to solving the riddle of its own existence. As my penetration of the past and future became indistinguishable, I became convinced that I was coming face to face with the ultimate origin and destiny of life, which were one and the same. I felt overwhelming, blissful certainty that there is one entity, one consciousness, playing all the parts of this pageant, and there is no end to this creative consciousness, only infinite transformations.

At the same time, my astonishment that anything exists at all became unbearably acute. Why? I kept asking. Why creation? Why something rather than nothing? Finally I found myself alone, a disembodied voice in the darkness, asking, Why? And I realized that there would be, could be, no answer, because only I existed; there was nothing, no one, to answer me. I felt overwhelmed by loneliness, and my ecstatic recognition of the improbability — no, impossibility — of my existence mutated into horror. I knew there was no reason for me to be. At any moment I might be swallowed up forever by this infinite darkness enveloping me. I might even bring about my own annihilation simply by imagining it. I created this world, and I could end it, forever. Recoiling from this confrontation with my own awful solitude and omnipotence, I felt myself dissolving, fracturing, fleeing back toward otherness, duality, multiplicity.

In the months after this nightmarish vision, I shaped it into a theodicy with gnostic overtones — call it gnosticism lite. I saw too much beauty and goodness in the world to condemn it and its creator outright, as the gnostics did. Perhaps the Demiurge is not evil, I conjectured, but only anxious and neurotic — and understandably so, given His existential plight. Creation — the multiplicity of the world — arises out of God's terrified confrontation with His own solitude, improbability, and potential mortality. Shunning His existential plight, God dissolves Himself into myriad selves, which compulsively seek but can never quite discover their true nature.

In the middle of the twentieth century, neo-Freudians proposed that as a result of horrific childhood trauma we may fragment into multiple personalities, which are ignorant or only dimly aware of each other's existence. Although multiple-personality disorder was all the rage in psychotherapeutic circles several decades ago, many psychiatrists now believe that the disorder is vanishingly rare, if it occurs at all. My 1981 experience nonetheless led me to suspect that we are all manifestations of God's multiple-personality disorder. At the same time, I realized that if anyone was disordered, I was.

Theodicies such as mine and those of the gnostics suggest that not all mystical visions fit neatly within the perennial philosophy, at least as Huston Smith explicates it. In recent decades, a growing number of religiously inclined scholars, including theologians, philosophers, and historians of religion, have complained that the perennial philosophy provides too simplistic a view of mysticism. Because they raise deep, troubling questions about our capacity for achieving universal truth, these scholars are sometimes called postmodernists (although like most so-called postmodernists, they refuse to accept that label). While they disagree about what mysticism means, postmodern religious scholars share a conviction that the perennial philosophy cannot possibly account for mystical experiences in all their strangeness, diversity, and darkness.

2

Attack of the Postmodernists

The cab ride from Midway Airport to Chicago's South Side passed through desolate inner-city landscapes. Looking for portents, I found them. On the side of a crumbling brick building someone had painted, "I love you . . . I love you . . . I love you. God." The taxi dropped me off at a malodorous hotel redeemed by its proximity to Lake Michigan, which on this gusty spring day was beryllium-hued, flecked with whitecaps. Walking along the lake toward the University of Chicago, I passed an ebony-skinned, snowy-haired man wearing white pajamas and pushing a shopping cart stuffed with suitcases. A professor of Sanskrit studies? A Hindu swami? A homeless schizophrenic?

My destination was a two-day conference titled "Mystics," sponsored by the University of Chicago's Divinity School. A flier promised that "leading scholars from the disciplines of history, literary theory, philosophy, and theology will share their current work" on mystics and mysticism. I hoped this meeting would give me a perspective on mysticism broader than, or at least different from, that of perennial philosophers such as Huston Smith, and my hope was fulfilled.

The contrast with the "Science and Consciousness" meeting in Albuquerque, where I had met Smith, was striking. None of the speakers here led the audience in guided meditation sessions or discussed their personal visionary experiences. (God forbid!) Unlike Smith, Aldous Huxley, and other perennialists, these academic scholars treated mysticism not as a universal human experience but as a literary phenomenon, a collection of "texts" requiring interpretation in the light of other texts.

The lectures recalled the postmodern literary theory I had been exposed to back in college. Postmodernists contend that poems, novels, even religious texts such as the Bible are best viewed not as expressions of timeless, transcendent truths but as products of the culture in which they are conceived. More radical theorists assert that all texts, even scientific ones, are not really about the world but only other texts. Language is ultimately a self-referential, hermetically sealed system, which reality always eludes.

I found these philosophical notions exhilarating when I first encountered them. At its best, postmodernism raises profound questions about the limits of language and knowledge. Over time, however, postmodernists reminded me of comedians who tell a clever joke, get a laugh, and then beat the joke to death, offering endless permutations to keep the laughs coming. The extreme postmodern outlook struck me as self-contradictory, even hypocritical. If you sincerely believe that truth is inexpressible, why bother speaking at all?

The Chicago meeting reminded me why I had found postmodernism so fascinating and so frustrating. Speakers kept alluding to negative theology, which is dedicated to describing and understanding a God who by definition cannot be described or understood. Negative theology is postmodernism distilled to its paradoxical essence, a conversation about the unspeakable. Whereas perennialists view ineffability as just one component of mystical visions, negative theology elevates it to the *sine qua non* of ultimate reality.

Some scholars in Chicago seemed to strive for ineffability in their presentations. Their speech bristled with postmodern buzzwords: hermeneutical, anagogic, cataphatic, semiotic, salvific narrative, sign and signifier, aporia. Often I had no clue what the speakers were driving at; they seemed engaged in missions scrutable only to themselves. An aphorism of the transpersonal philosopher Ken Wilber came to mind: "Academic religion is the killing jar of Spirit." During one particularly opaque talk, my eyes fell shut, and I drifted into unknowing until my own snores brought me back to the fallen world again.

Occasionally, however, vivid one-liners popped out of even the densest lectures: "God cannot be God unless He first becomes the devil." "Ecstasy is engendered by confronting the real in all its horror." "I am because I will die, and I am human because I know I will die." "If God could be known, this would be the best evidence that this is not God." "God is not only sleeping, he is snoring." An Australian literary theorist served up the best joke of the meeting. He opened his talk by confessing that he had mistakenly shown up for the conference a day early. When he walked into the empty auditorium, he thought to himself, "This is taking negative theology a bit too far."

Gradually, I discerned behind the clotted academese the same spiritual yearning I had sensed in Albuquerque. Here the desire for transcendence was if anything more poignant, because to these scholars mystical knowledge was not just a Zen session or mescaline capsule away; it was impossibly distant. The speakers were also acutely aware of the insufficiencies of language, their chosen path toward truth. They took seriously what Dionysius the Areopagite, a mysterious sixth-century monk, once said of the mystical path: "The more it climbs, the more language falters, and when it has passed up and beyond the ascent, it will turn silent completely, since it will finally be at one with Him Who is indescribable."

"Mystical literature is that which contests its own possibility," the Australian literary theorist Kevin Hart lamented. "It is through the fissures of our discourse that the mystical is seen." The French philosopher Jean-Luc Marion agreed that mysticism represents language's greatest challenge; it evokes Ludwig Wittgenstein's cryptic aphorism: "Whereof one cannot speak, thereof one must be silent." But Marion believed that poetry and literature, modes of discourse that "require endless interpretations," can help us express the inexpressible. Kevin Hart retorted that if you can't talk about the mystical, you can't "whistle" about it either. In other words, he explained, art and literature are inadequate, too.

The meeting also revealed the dizzying diversity of mysticism just within the Judeo-Christian tradition. Several speakers discussed Meister Eckhart and Dionysius the Areopagite, who are seminal figures of negative theology. Perennialists often invoke Eckhart and Dionysius, because their descriptions of the divine as "nothingness" or a "desert" resemble Buddhist and Hindu references to the void underlying all things. These are the apophatic visions that Huston Smith and other perennialists claim reveal the supreme mystical truth.

But lecturers recalled other mystics whose visions do not fit neatly into the perennialist scheme. There was the fifteenth-century peasant Joan of Arc, who heard angelic voices commanding her to lead French troops against the British. A century later Teresa of Ávila lapsed into rapturous swooning fits in which she felt herself penetrated by shafts of divine love. For the twentieth-century French philosopher Georges Bataille, mystical awareness was a kind of ecstatic despair, which he cultivated by staring at photographs of torture victims.

These talks hinted at a mysticism darker and more diverse, wilder and more resistant to explanation than the mysticism of the perennialists. So did the suggestion of several speakers that mystical experiences can reveal not God's presence but His absence. The Jewish theologian Michael Fishbane addressed this issue in a talk on the Kabalist doctrine of *tsimtsum*, which suggests that in the act of creating us, God withdrew from us into some infinitely distant empyrean. We can only know Him because of His otherness, His apartness. According to this doctrine, Fishbane said glumly, mystical knowledge represents not blissful union with God but a "radical form of alienation" from Him.*

A similar notion emerged in the lecture of Bernard McGinn, a theologian at the University of Chicago, author of a multivolume history of Christian mysticism and chief organizer of the "Mystics" conference. McGinn commanded the respect of other speakers; many referred — and deferred — to his work in their talks. He was professorial in demeanor and attire; clad in a blue blazer and bowtie, he had thinning white hair and a white beard. There was a marked contrast between the matter-of-fact delivery of his lecture and its nightmarish substance.

When Dionysius and Eckhart refer to God, McGinn remarked, their descriptions are stripped not only of sensory content but also of emotion; they seem to be perceiving literally nothing at all. Eckhart, for example, described the encounter with God as "the cessation of all experience." For other mystics, McGinn pointed out, transcendence can be pure anguish. They are gripped by "horrible terror," "contemplative agoraphobia," a sense of "dereliction." God seems to have vanished, possibly forever.

McGinn described a thirteenth-century nun named Angela who spent two anguished years convinced that God had abandoned her. She compared herself to someone who hangs by the neck but never dies. Finally she entered a state that she described as "beyond love and knowing," and she heard the words "I am you, and you are I." A nun called Margaret the Cripple also oscillated between periods of agonizing estrangement from

*Spelling out the disturbing implications of *tsimtsum*, Arthur Green, a scholar of Jewish mysticism at Brandeis University, once wrote ("The Aleph-Bet of Creation: Jewish Mysticism for Beginners," *Tikkun*, vol. 7, no. 4, p. 72): "To say that I exist thanks only to God's absence is to make me a being who thrives on alienation, who needs to be far from the One at least as much as he needs to be near. Will I then seek darkness as much as light? Evil as much as good? How dare I then speak of union?" For lucid discussions of *tsimtsum*, *shekhinah*, *tikkun*, and other Kabalist concepts, see also Green's book *Seek My Face, Speak My Name* (Jason Aronson, Northvale, New Jersey, 1994).

God and rapturous communion with Him. Because they never lost their faith in God, McGinn said, these female mystics endured and even embraced their periods of despair. "They seek it out," he said, "in the confidence that it is only through embracing God's absence, even hell itself, that God can somehow be attained. True consolation is in desolation."

The postmodern outlook is often described as nihilistic, hostile to any type of belief. But for some scholars — who could be called, oxymoronically, postmodern believers — postmodernism serves the opposite function: By questioning the concept of universal truth, postmodern believers seek to liberate us to believe what we choose, whether Hinduism, Kabalism, gnosticism, or existentialism. The catch is that you cannot claim your belief system is binding for anyone else.

When I spoke to Bernard McGinn after his lecture, I concluded that he is a postmodern believer. He is a Roman Catholic, and his mysticism scholarship, he said, is an expression of his personal faith. McGinn is also an anti-perennialist. The perennial philosophy, he complained, strips Christian mysticism of precisely those religious distinctions that he as a Catholic finds most meaningful. Examining different mystical traditions for commonalities is a worthy endeavor, McGinn said; in fact, he has done comparative studies of Christian, Jewish, and Islamic mysticism. But he disapproved of perennialist attempts to equate Meister Eckhart with Zen masters such as Dogen, or Christian salvation with Buddhist enlightenment.

McGinn's chief objection to the perennial philosophy is that it does not do justice to the vast diversity of mystical phenomena. "One of the important things about mysticism," McGinn said, "is not to restrict it to some particular kind of experience." McGinn pointed out that perennialists commonly describe mystical experiences as some sort of union with God.

Actually, McGinn said, few Christian mystics report such unitive experiences; Christian mysticism might be more aptly defined as an encounter with the "presence" of God, who remains apart. But even this definition is too restrictive, because many mystics sense God's *absence.* To these mystics, McGinn commented in his magnum opus, *The Foundations of Mysticism,* God "becomes a possibility only when the many false gods (even the God of religion) have vanished, and the frightening abyss of total nothingness is confronted." These are the sorts of hellish visions that McGinn had discussed in his lecture in Chicago.

McGinn was careful to distinguish the medieval nuns who had those visions from the heretical Christians known as gnostics. The Catholic nuns attributed their alienation from God to their own shortcomings, whereas the gnostics concluded that there is something wrong with God, or rather, the Demiurge who created this world. When I asked McGinn if he had any personal acquaintance with hellish or heavenly mystical states, he chuckled uneasily and replied, "I tend to try to stay away from that."

Steven Katz looks through a glass darkly

One reason I found McGinn's lecture so gripping was that it reminded me of the hellish experience I underwent in 1981, when I was an undergraduate at Columbia University. A few months after that trip, desperate to understand what had happened to me, I took a class titled "Mysticism." We read tracts by the ninth-century Hindu sage Shankara, Meister Eckhart, Teresa of Ávila, and other mystics, plus scholarly analyses by William James, the philosopher Ninian Smart, and others. Our most recent text was *Mysticism and Philosophical Analysis,* a collection of essays by religious scholars published in 1978. The book was edited by the philosopher of religion Steven Katz, who also contributed two essays.

What I did not fully appreciate at the time was that Katz's book was an all-out assault on the perennial philosophy. In fact, the underlying premise of the Chicago "Mystics" meeting — the assumption that mysticism is best seen as a textual rather than experiential phenomenon, which must be understood within its particular religious context — echoed what Katz has been arguing for more than twenty years.

Katz established his sardonic tone in his introduction, where he accused perennialists of being evangelists rather than impartial scholars. Perennialists exhibit "strong biases," Katz asserted, that "significantly diminish the value of their results." Perennialists believe that identifying the common threads of all religions will help them coexist more harmoniously; they also hope that, in an era when science has undermined the basis for faith, mystical experience might give us the truth and meaning we crave. "The cure for the anomie created by our revolutionary contemporary awareness of man's finitude, mortality, and freedom," Katz wrote, "was now sought in a renewed, immediate, non-critical, largely non-cognitive contact with the mystical depths of Being itself (whatever that is!)."

Katz identified three main versions of the perennial philosophy: The simplest holds that all mystical experiences are essentially the same, and that descriptions of them by different mystics "reflect an underlying similarity which transcends cultural or religious diversity." A slightly more sophisticated version agrees that all mystical experiences are the same but grants that mystics' descriptions of their experiences vary according to their cultural background. The most sophisticated perennialist position divides mystical experiences into a few basic categories that transcend cultural boundaries but are described in culturally bound language.

Katz contended that *all* of these positions are flawed, and that "there is no *philosophia perennis,* Huxley and many others notwithstanding." Katz spelled out his own assumption in a single sentence, which he italicized for emphasis: "*There are* NO *pure (i.e., unmediated) experiences.*" There is no universal mystical truth apprehended by Christians, Jews, Buddhists, and Hindus alike; mystical experiences are enormously diverse, and they invariably reflect each mystic's particular culture and personality.

Katz's critique of perennialism, which he expanded upon in two subsequent books, had a tremendous impact on scholarly discussions of mysticism. Whereas perennialists had once dominated theological and philosophical discussions of mysticism, they now became an embattled minority. It didn't hurt that Katz's viewpoint dovetailed with the broader trend in academic scholarship toward postmodernism.

Rereading Katz almost twenty years after my first exposure to him, I found myself both impressed and unsettled by his no-unmediated-experiences argument. It echoed the philosopher Thomas Kuhn's proposal that all scientific theories are to some extent products of a particular mindset, or "paradigm," that implicitly constrains the questions and answers that scientists formulate. Kuhn's philosophy implies that science cannot achieve absolute truth; as paradigms change — which they inevitably do — so does the so-called truth.

Katz's stance also reminded me of Timothy Leary's assertion that psychedelic experiences are influenced by "set and setting" — that is, the prior mindset of the drug taker and the setting of the drug session.* Katz seemed to believe that set and setting not only influence but virtually determine mystical experiences, that universality is a chimera. In other words, Huston

*According to Thomas Roberts, a psychologist at Northern Illinois University whom I met at the "Mystics" meeting in Chicago, the phrase "set and setting" first appeared in print in "Reactions to Psilocybin Administered in a Supportive Environment," by Timothy Leary, George H. Litwin, and Ralph Metzner (*Journal of Nervous and Mental Disease,* vol. 137, no. 6, 1963, pp. 561–573).

Smith and Aldous Huxley did not really glimpse some transcendent reality when they consumed mescaline; their visions were merely byproducts of their hopes and fears and cultural indoctrination.

I wondered about Katz's motives. Was he trying to cast doubt on all mystical experiences and on all religious and spiritual propositions? Was he covertly defending a particular religious outlook that does not fit into the perennial philosophy? Did Katz himself have a mystical vision at odds with that philosophy? Or was he denigrating mystical experiences because he never had one?

On a brilliantly sunny fall day, I met Katz in his office in Boston University's School of Theology, where he heads the Department of Jewish Studies. He was as combative in person as on the page. A corpulent man with thinning white hair, he wore maroon suspenders that framed his belligerently convex chest and belly. Wit leavened his pugnacity. He delivered his jokes deadpan, and he seemed genuinely unconcerned with whether I appreciated his humor.

Commenting on the moral dimension of Buddhism, he noted that Buddhists "are concerned with *all* living things, whether low-level living things like sociologists or high-level living things like philosophers of religion." When I asked if he thought his opponents would ever be persuaded by his arguments, Katz shrugged. "We have institutions that are full of people who are not persuadable. From Congress to other kinds of institutions, right?" Katz was raised in northern New Jersey, and his Jersey accent made his denunciations sound even more barbed. "Huston's a *nice* man," Katz said, his voice oozing condescension. "He's just not very" — Katz paused for maximum effect — "strong."

I told Katz that it helps me, when trying to understand someone's attitude toward mysticism, to learn something about his background. Katz was one step ahead of me. "Oh, I had a mystical experience!" he said, bugging his eyes out and fluttering his hands like a Holy Roller preacher. "God said to me, 'Go forth!' That the kind of evidence you want? It's a strictly academic thing!"

Personal experiences are "absolutely irrelevant" to the debate over perennialism, Katz declared. "All I have available of the authentic or lack of authentic mystical experiences of Eckhart, Teresa of Ávila, John of the Cross, Shankara, is what they have written. I have no access to their minds. So personal experience doesn't count very much in this." Katz had long ago vowed never to discuss his personal religious convictions or experiences.

Katz did assure me, however, that he had never taken psychedelic drugs, nor did he have any intention of doing so. He scoffed at the claim that drugs can induce genuine mystical experiences. He was unimpressed with the results of the Good Friday experiment, which took place just down the street from his office. Those who think they have had mystical experiences after ingesting drugs "*didn't* have mystical experiences. They had *drug* experiences," Katz said. "All you are doing when you're taking a drug is experiencing your own consciousness."

But that could be said of any method for inducing mystical experiences, I replied, including meditation, yoga, prayer, and fasting.* "Those are different," Katz responded. "The object is not to change brain chemistry but to put you in touch with metaphysical realities." Most of those who consume psychedelics seek not transcendent truths but merely sensory titillation, he said. "People don't take these drugs for metaphysical reasons." You're wrong, I said. People take drugs for many reasons, just as they practice yoga for many reasons: to get exercise, to relieve anxiety, to escape humdrum reality, or to find God.

Katz shook his head at my obtuseness. What makes mystical experiences so fascinating, he said, is the possibility that they represent glimpses of ultimate reality, or God. "They seem to confirm experientially what most people know only by propositions about the nature of these ultimate realities. Drug experiences are irrelevant to all of that, because all they do is tell you about what is going on in your mind," Katz said. "I mean, I give you a mushroom, you have an experience. You think that immediately is a contact with *God?* The portal to heaven is through a *mushroom?*"

Why is a mushroom a more absurd route to God, I asked, than kundalini yoga or flagellation or the study of Kabala? I added that William James put this issue to rest in *The Varieties of Religious Experience:* Ultimately, James wrote, it doesn't matter whether a mystical experience stems

---

*Huxley discussed the physiological effects of spiritual practices such as "yogic breathing" and chanting (increase carbon dioxide in the blood), fasting (decreases blood sugar), and flagellation (reduces histamines and adrenaline) in *Heaven and Hell* (copublished with *The Doors of Perception,* Perennial Library, Harper & Row, New York, 1990). He declared that "*all* our experiences are chemically conditioned, and if we imagine that some of them are purely 'spiritual' . . . it is merely because we have never troubled to investigate the internal chemical environment at the moment of their occurrence" (p. 155). Given mescaline and other new-and-improved mystical technologies, Huxley added, it is "senseless" for an "aspiring mystic" to fall back on primitive techniques such as fasting and flagellation (p. 156).

from a drug, a prayer, a divine intervention, or a brain disease; the experience must be evaluated on its own merits.

"Do you know the story of William James taking drugs?" Katz retorted. "He said to his students, 'When I am under the influence of the drug, I want you all to write down everything I say.' And he took this drug, ethyl chloride, I think it was. When he came out, he was very anxious, interested. He asked, 'What did I say? What did I say?' And his students said" — Katz droned, as if in a trance — " 'The . . . whole . . . world . . . smells . . . like . . . ethyl . . . chloride.' "

I had to laugh. We agreed to disagree and moved on. Katz said that he first gave careful consideration to the perennial philosophy in the early 1970s, when he taught a course on the history of religion at Dartmouth. Reading Huston Smith and other perennialists, Katz rebelled at what he felt was their distorted presentation of Jewish mysticism. Delving more deeply into other mystical traditions, Katz concluded that they, too, had been misrepresented to make them more mutually compatible.

Contrary to perennialist claims, Eastern and Western mysticism are "altogether different" in their "form and structure and character," Katz said. Within each religion there are further divisions. "There are Theravada Buddhists and Mahayana Buddhists and Pure Land Buddhists and Zen Buddhists. And Buddhism in China is different from Buddhism in Burma, and Buddhism in Burma is different from Buddhism in Thailand." Each of these traditions seeks and interprets mystical experiences in different ways.

Katz rejected the charge of some of his critics that his anti-perennialism is a covert defense of his own faith, Judaism.* Although he is an Orthodox Jew, Katz said, he does not consider himself devoutly religious. To be

---

*One perennialist who suspects that Katz's anti-perennialism masks a defense of Judaism is Robert Forman, a religious scholar and veteran practitioner of Transcendental Meditation. "The upshot of [Katz's] argument," Forman told me, "is that there's something special about Judaism, and to analyze Judaism as if it's just another perennial philosophy, or just doing mysticism like everybody else, kind of undercuts the authentic idiosyncrasy of Judaism." Forman has criticized Katz in several books, including a collection of essays that he edited, *The Innate Capacity* (Oxford University Press, New York, 1998). Ironically, one article in *The Innate Capacity* seemed to support Katz's view of mysticism rather than Forman's. In "The Swami and the Rorschach," Diane Jonte-Pace described responses of various "spiritual masters" — including a Hindu swami, Buddhist adepts, and an Apache shaman — to Rorschach blots. The responses "differed considerably; there was clearly no universalism of content," Jonte-Pace concluded (p. 142). An Apache shaman was reminded of natural phenomena such as lightning and the changes of season; a Hindu swami saw the union of the divine forces Shiva and Shakti; and a Buddhist sage saw the human mind trapped in the suffering caused by egotism.

sure, he added, some perennialists have been dismissive or even hostile toward Judaism. Joseph Campbell, the celebrated mythologist, was a case in point. After Campbell's death in 1987, reports emerged that he privately expressed distaste for Jews and Judaism.

"Now, he may have been anti-Semitic because he just didn't like Jews," Katz said. But Campbell's negative attitude toward Judaism, he suspected, was a logical consequence of his perennialist outlook. "Perennialists, because they don't like differences, play down law and dogma and ritual. And Judaism," Katz added drily, "is not unknown to have law and dogma and ritual." Campbell "didn't like Western religions in general. He preferred Asian religions. But of the Western religions, the one he liked least was Judaism, because it was most unlike the things he liked."

Actually, Katz said, the version of Eastern religion favored by Campbell and other Western scholars only faintly resembles the originals. "It's the Asian religion of Americans and Californians," he continued. "It's not real Buddhism or real Hinduism." When Americans go to genuine Hindu or Buddhist monasteries rather than those that cater to Westerners, "they leave very soon; they're not happy," Katz said. "So this is like the old melting pot in America. It's a melting pot, but you have to be melted down to be a WASP."

Katz denied that he is a radical postmodernist who considers truth to be unattainable. Nor does he believe that mystical experiences are entirely fabricated from mystics' prior beliefs and expectations. "That's *not* what I say. I say there is a dialectic between our environment and our experience." Katz called his position contextualism. "What I do is look for context, and I try to give the broadest possible phenomenological account of the context."

But he implied that it may be virtually impossible for us to find any truth that transcends our personal context. Although less dismissive of meditation than of psychedelic drugs, Katz rejected the claim of some meditation enthusiasts that meditation "deconditions" the mind, cleansing it of preconceptions so that they can perceive ultimate reality unfiltered. Meditation merely *re*conditions the mind, Katz said, substituting one set of expectations and beliefs for another.

Don't the questions you raise about mystical experience, I asked, endanger all theological systems, including Judaism? Katz stared at me and slowly, with theatrical befuddlement, repeated my question. I tried again, saying that his emphasis on the role of expectation and education in shaping mystical experiences could undermine religious belief in general. Katz grimaced, shaking his head. "The interpretation of my account — that it would lead somehow to reductionism, or undermining of religious belief — I don't see why that's the case."

Far from undermining all religious faith, Katz said, his approach fosters open-mindedness and tolerance. "My view starts with the legitimacy of all metaphysical claims. I make no claims about absolute reality. I don't say it's God who's misunderstood by the Buddhists. I don't say it's nirvana that's misunderstood by the West," he said. "I *never* do that. That seems to me to be chutzpah, metaphysical chutzpah."

It is perennialism that threatens religious faith, Katz argued, by forcing all religious and mystical experiences into a procrustean bed. Perennialists "are like the colonialists who come from Spain to convert all the Indians because they are all savages," Katz said, heating up. Perennialists "think they're being ecumenical; they're saying everybody has the same belief. But they are doing injustice to all the people who say, 'I'm *not* believing like you do.'"

I accused Katz of underestimating just how disturbing his outlook is. The perennial philosophy attempts to solve a problem: Each of the world's traditions claims to present The Truth about reality. So which tradition is right? One rational response might be that none is right, all are wrong. This view leads to atheism, or nihilism. The perennialists try to avoid this conclusion by finding commonalities among all religions and all mystical visions. Katz puts everything back to square one. How can we determine which religious claims are valid and which are not?

"We don't know," Katz replied quietly, his combativeness subsiding. "There doesn't have to be *the* right way. There can be *a* right way." Perhaps there is one absolute reality, Katz said, which each of us can only glimpse through a glass darkly. "All of us see only an aspect, only an attribute, only a partial vision. And ultimate reality, by its very nature, escapes us, because it is ultimate reality, and as human beings we are partial observers. It doesn't mean we *don't* see."

It would be "nice," he added, "if we shared one view, and we didn't fight about religion, and God made it clear to us that *this* is the way to salvation. But whether He is just a joker, and likes the sport, or it's all an enormous mistake, we have all these religions, and they are all different."

Like the Catholic theologian Bernard McGinn, Katz is a postmodern believer. I believe Katz believes that by attacking the perennialist imperialists he is protecting the autonomy of all religious traditions, not just Judaism. But Katz's contextualism (to use the term he prefers) can be used to challenge any intellectual position, including his own. Katz was raised in an Orthodox Jewish family in an Orthodox community. At Cambridge University, where he received his doctorate, he studied under Gershom Scholem, the great scholar of Jewish mysticism. Scholem was a staunch anti-perennialist; he once asserted that "there is no mysticism as such, there is only the mysticism of a particular religious system, Christian, Islamic, Jewish mysticism, and so on." In other words, Katz's contextualism reflects his personal context.

The fact is, Katz's philosophy is more compatible with atheism than with belief. It implies that mysticism can tell us nothing about ultimate matters; mystics make divergent, inconsistent, and even contradictory claims about the nature of ultimate reality, which reflect mystics' prior conditioning. We have no way to tell which claims, if any, are valid. The logical conclusion would be that all mystical visions are illusions.

The scandal of particularity

In a 1987 essay titled "Is There a Perennial Philosophy?" Huston Smith challenged the contention of Katz and other scholars that human experiences, beliefs, and behavior can be understood only within their particular sociohistorical context. When pushed too far, Smith warned, this postmodern outlook culminates in a kind of "cultural solipsism," which "renders unintelligible the ways and degree to which we can and do communicate, understand, and, yes, even experience cross-culturally."

Smith's point is well taken. We must share certain universal experiences in spite of our circumstantial differences, if only because we share the same biology. Descriptions of hunger and thirst and sexual desire may vary from person to person and culture to culture, but does that mean these basic biological drives are products of our social conditioning? Certain aspects of the external world, too, transcend context. If Aristotle's descriptions of the moon differ from those of Carl Sagan, does that mean they didn't see the same thing? Katz's contextualism implies as much.

Mysticism is stranger and more complicated than perennial philosophers often imply; scholars like Katz and McGinn have made this point convincingly. But their stance — whether called postmodern or contextualist — is also too indiscriminating. They argue, in effect, that all mystical revelations are unprovable, hence they cannot be distilled into a single, coherent worldview such as the perennial philosophy. All mystical visions may be unprovable in an absolute sense, but surely some are more demonstrably false — and potentially harmful — than others.

The question is, how do we interpret a religious or mystical vision? How do we integrate it into our preexisting view of the world? Do we consider our vision to be literally or only metaphorically true? Do we see it, for example, as a symptom of urges that we keep bottled up in our individual unconscious (as Freudians might say), as a message from some collective human psyche (as Jungians would argue), or as a sign from God Himself?

Huston Smith has provided me with one crucial criterion for judging religious visions, although both Smith and the anti-perennialists Katz and McGinn would probably disapprove of how I employ it. In *The World's Religions,* Smith raised an issue called "the scandal of particularity." This theological phrase, he explained, refers to "the doctrine that God's doings can focus like a burning glass on particular times, places, and people(s) — in the interest, to be sure, of intentions that embrace human beings universally." The essence of the doctrine is that God plays favorites.

In its crudest form, this doctrine manifests itself in the declarations of professional athletes on national television that the Almighty helped them triumph over their opponents. Not just the team's *belief* in the Man Upstairs but that Man Himself helped the quarterback of the Jacksonville Jaguars complete that key pass on the last play of the game, propelling his team to victory over the evil Denver Broncos. A miracle!

It's easy to dismiss the notion of a God who fixes sporting events. But all religions assert in one way or another that God plays favorites. The founders of the great Western faiths — Abraham, Jesus, Muhammad — were supposedly singled out by God for special treatment. The religions they founded proclaim that those of us who act in a certain way and believe certain things will also be divinely favored. Buddhism and Hinduism, too, promise that good behavior will be rewarded and bad punished through the abstract mechanism of karma. The pious monk goes to nirvana after death, and the nasty industrialist is reincarnated as a cockroach.

All these schemes ask us to believe in some sort of supernatural moral accountant who, like Santa Claus, keeps tabs on our naughtiness and niceness in order to determine our fate in the afterlife. As William James commented: "Any God who, on the one hand, can care to keep a pedantically minute account of individual shortcomings, and on the other hand can feel such partialities, and load particular creatures with such insipid marks of favor, is too small-minded a God for our favor."

Huston Smith wrote *The World's Religions* not to criticize religions but to promote understanding and respect for them. Hence he did not dwell on the scandal of particularity beyond a brief acknowledgment that it violates our "principles of impartiality and fair play." To my mind, the scandal of particularity is the root of all religious evil. The conviction of certain individuals and peoples that they are divinely chosen leads to religious self-righteousness, fanaticism, intolerance.

Also, what kind of God would play favorites? If such a God exists, do we really want to worship Him? Wouldn't that mean that our faith is just a form of self-interest? A God who physically intervenes in human affairs — whether by parting the Red Sea or guiding a football to a receiver — also contradicts science's hard-won discovery that the world is governed by impartial laws and not by divine caprice. For all these reasons, I am inclined to reject as delusional any mystical vision — or spiritual doctrine derived thereof — tainted by the scandal of particularity. All of us must be God's sons and daughters, or none of us. All are chosen — or damned — or none.

Even with these restrictions, we still have a confusing range of revelations to ponder, from the heavenly to the hellish. Is there any way of judging, however tentatively, which are true and which are delusional? At the "Mystics" conference in Chicago, several speakers noted that medieval mystics were often suspected of being inspired not by God but by the devil. Nowadays, the issue is more likely to be stated in psychiatric terms: Could the mystic's vision be a delusion stemming from psychopathology?

Incredibly, not a single speaker in Chicago raised this question, but one audience member did. After Bernard McGinn gave his talk on the harrowing mystical experiences of the medieval nuns Angela and Margaret, Thomas Roberts, a psychologist at Northern Illinois University, rose from the audience to comment. Because I had met Roberts earlier, I knew he had been fascinated by mysticism ever since a mystical psychedelic encounter in 1971. He had recently helped create the Council on Spiritual Practices, a San Francisco–based organization that advocates entheogenic methods for enhancing spirituality. Roberts had hoped to find insights into his entheogenic experiences at this conference, but he had soon concluded that none of these scholars really knew anything about mysticism. He complained to me during a coffee break that a eunuch, while he can observe and write about the sexual behavior of others, obviously has no real insight into the subjective *experience* of sex. The so-called experts at this symposium, Roberts grumbled, are "mystical eunuchs."

Although he did not repeat this assessment to Bernard McGinn, Roberts made it clear that he found McGinn's treatment of mysticism unsatisfying. He asked if McGinn had considered nonreligious explanations for the nightmarish experiences of these medieval female mystics. For example, their terror and sense of divine abandonment might have been reactions to the bubonic plague then ravaging Europe. The women might have been afflicted with what would now be diagnosed as schizophrenia or manic depression. Or perhaps they had been poisoned by the ergot molds that often infected grains in medieval Europe; consumption of these molds can produce psychotic symptoms similar to those induced by LSD.

Roberts was offering a perspective conspicuous by its absence at the Chicago meeting — and, more generally, in discussions of mysticism among religious scholars. By examining historical cases of mysticism in the light of modern psychology and psychiatry, Roberts was suggesting, we might be able to make distinctions between them. We might even determine which are genuinely timeless revelations that transcend their particular context and which are products of pathology or cultural idiosyncrasy.

McGinn would have none of it. He granted that there could be psychological or physiological explanations for some mystics' experiences. But he believed that these experiences are best understood within their contemporary religious context, on their own terms. "The idea that mysticism can be extracted from its time and place I think is a mistake," McGinn told Roberts.

The crazy-wisdom conundrum

Ironically, Bernard McGinn himself has suggested that religious scholars trying to comprehend mysticism might have much to learn from psychology, psychiatry, and other mind sciences. In *The Foundations of Mysticism,* McGinn noted that not all mind scientists dismiss mystical experiences as "mere regressions or repressions of sexuality," as Sigmund Freud did. McGinn was encouraged by the fact that some "recent psychological investigators," the psychiatrists Stanislav Grof and Arthur Deikman among them, have treated mysticism in a sympathetic manner instead of reducing it to psychopathology.

In fact, over the past half century an eclectic band of psychiatrists, psychologists, and other spiritually inclined intellectuals have formed a field, called transpersonal psychology, that embraces mysticism and spirituality. Distinguishing genuine, healthy mysticism from madness is a central concern for transpersonal psychology. A century ago, the proto-transpersonalist Richard Bucke warned, in *Cosmic Consciousness,* that as consciousness expands, so does the risk of insanity. Over the next hundred years, Bucke's warning seemed to be borne out by the antics of some prominent twentieth-century gurus, professional mystics who claim to be enlightened and hold out the promise of enlightenment to devotees.

G. I. Gurdjieff, who early in the last century attracted celebrity followers such as Aldous Huxley, Georgia O'Keeffe, and Frank Lloyd Wright, was obsessed with the moon. The moon, he proclaimed, was absorbing energy released by the death of living organisms on earth and would eventually turn into a fertile, earthlike planet in its own right. The self-proclaimed "perfect master" Meher Baba built an international following before he died in 1969, even though for more than forty years he had refused to speak and communicated only with a chalkboard. Another Indian swami, Nityananda, was known to sit naked on piles of excrement, smear it on his body, and even eat it on occasion.

Other gurus crossed the border from eccentricity into sociopathy. Bhagwan Shree Rajneesh attracted tens of thousands of followers in the 1980s while advocating sexual promiscuity and violence as avenues to transcendence. A community that he founded in Antelope, Oregon, dissolved — and Rajneesh was deported back to India — after high-ranking devotees were charged with various crimes, including poisoning Antelope officials opposed to the community's plans for expansion. Shoko Asahara, leader of Japan's Aum Shinrikyo sect, was imprisoned in 1995 after his followers released nerve gas in Tokyo subways, killing twelve people and injuring thousands. Asahara reportedly hoped to initiate an apocalyptic war that he had foreseen in a vision.

Chögyam Trungpa, who introduced Tibetan Buddhism to the West in the 1970s and founded the Naropa Center in Boulder, Colorado, had sex with male and female followers and showed up for lectures hours late and drunk. When the poet W. S. Merwin and his wife visited Naropa in 1975 and declined to join a drunken party, Trungpa and his disciples kicked in the door of their room and forcibly stripped off their clothes. Before he died of acute alcoholism in 1987, Trungpa appointed an American acolyte named Thomas Rich, also known as Osel Tendzin, as his successor. Rich, a married father of four, died of AIDS in 1990 amid published reports that he had had unprotected sex with male and female students without telling them of his illness.

Are gurus like these simply psychopaths? Not necessarily, according to Georg Feuerstein, an authority on Eastern mysticism and a contributor to transpersonal psychology. In his 1991 book *Holy Madness,* he reviewed the history of dozens of scandalous gurus, including all those mentioned above except for Asahara. Trying to put these cases in perspective, Feuerstein argued that all great spiritual teachers, including Jesus and Buddha, challenged social norms in ways that could have been judged insane. Throughout the history of spirituality, moreover, some spiritual adepts have acted in especially unconventional, even shocking ways. This behavior is called holy madness, or crazy wisdom.

Although generally associated with Hinduism and Buddhism, crazy wisdom has cropped up in Western faiths, too. After Saul became Saint Paul, he preached that a true Christian must "become a fool that he may become wise." Paul's words inspired a Christian sect called Fools for Christ's Sake, members of which lived as homeless and sometimes naked nomads. Sufi mystics called Malamatis embraced "the path of blame"; they deliberately behaved badly to arouse society's loathing and hence, they believed, God's love.

Feuerstein deplored the abusive behavior of certain modern gurus, and he granted that some may actually be madmen or charlatans. But others may be genuine mystics who transgress for our benefit, to show us how to transcend our moral and cognitive limits and know true, mystical liberation. Feuerstein, who was once a devotee of a crazy-wisdom adept called Da Free John, said that determining whether a guru is actually enlightened is not an easy matter. "We cannot know the inner man or woman unless we happen to be great *yogins* ourselves." That is, you must be enlightened to know for sure whether someone else is.

Who would dare to claim such expertise? Certainly none of the scholars at the "Mystics" meeting in Chicago, nor Huston Smith, who views enlightenment as an unattainable ideal. But there is one major figure in the field of transpersonal psychology who says he has achieved "constant nondual awareness," a mystical consciousness so powerful that it persists even through deep sleep. He claims, in other words, to be enlightened.

3
The Weightlifting Bodhisattva

On August 7, 2000, in the midst of the U.S. presidential race, an op-ed piece in the *New York Times* revealed that the democratic candidate, Al Gore, enjoyed reading the works of the transpersonal maven Ken Wilber. The article identified Wilber as a disciple of New Age mogul Deepak Chopra. Ten days later the *Times* printed a letter from Chopra correcting the record. "I am Ken Wilber's disciple," Chopra revealed, "not the other way around." Unlike the scholars I heard at the "Mystics" conference in Chicago, Wilber has no formal academic affiliation or a degree in religious studies. And yet he is almost certainly as knowledgeable about the history of mysticism as any academic scholar. He has also steeped himself in spiritual practices of East and West alike, including Zen, tantric yoga, Sufism, and Kabalism.

Wilber was a graduate student in biochemistry at the University of Nebraska in the early 1970s when he started meditating and devouring books on mysticism and related topics. He dropped out of graduate school to write a massive book outlining convergences between Eastern mystical traditions and Western philosophy and psychology. Since *The Spectrum of Consciousness* was published in 1977, Wilber has churned out more than a dozen books, and he has earned an international reputation as a thinking person's spiritual author.

Like Huston Smith, who helped him get his first book published and has championed his work ever since, Wilber is a staunch proponent of the perennial philosophy. But unlike Smith, Wilber views science as an ally in the perennialist cause. Nor does Wilber attempt to preserve religions in all their diverse glory, as Smith does. Quite the contrary. Wilber applauds the fact that science is stripping us of our "infantile," prescientific beliefs. He deplores New Agers who conflate mysticism with irrationality and a retreat into primitive superstition. A mature mysticism incorporates rationality and then transcends it, Wilber insists; our culture's ultimate goal should be fusing "the Enlightenment of the East with the Enlightenment of the West."

Here is another point where Wilber's worldview diverges from Huston Smith's: Whereas Smith downplays enlightenment and even doubts whether it is completely achievable, Wilber makes enlightenment the be-all and end-all of existence. Wilber dismisses the contention of anti-perennialists such as Steven Katz that because mystics' visions often reflect their personality, background, and cultural context, we cannot extract any universal truths from them. Wilber compares this argument to that of post-modern philosophers who claim that all so-called truths, including scientific ones, are actually just inventions that serve human needs and change as our needs change. This position is absurd, Wilber says; scientists have discovered many absolute truths, and so have mystics.

Wilber is unusual among modern spiritual writers in rejecting attempts to link spirituality to physics, which, as he points out, addresses only the basic level of material reality. The disciplines relevant to mysticism, he argues, are psychology, psychiatry, and, to the extent that it focuses on the mind, philosophy. The seemingly contradictory views of the mind held by thinkers as diverse as Buddha, Plato, Sigmund Freud, Carl Jung, Jean Piaget, and Erik Erikson can be reconciled by realizing that each addresses different levels of consciousness.

Scientists, among them Freud and Piaget, usually stop well short of the highest levels of consciousness, whereas some mystical philosophers chart only the loftiest regions and neglect the lower ones. But when all the various models of the mind are put together, Wilber contends, they form a clear-cut picture: We begin life as merely physical beings, no more aware than bacteria. Even as children, we are still self-absorbed creatures. As we develop emotionally, intellectually, and spiritually, we become aware of an ever-wider sphere of existence. Our awareness matures into an empathy for and identification with all of humanity, all of nature, and ultimately all of existence, including the eternal void from which all things come.

For someone who proclaims the unity underlying reality, Wilber is fond of slicing and dicing it. He starts with the Great Chain of Being, a hierarchical model of reality posited by mystical thinkers ranging from the third-century Neoplatonist Plotinus to the twentieth-century Indian yogi

Sri Aurobindo. The Great Chain rises from mere matter up through the lower stages of mind and culminates in Godhead. Wilber prefers to call it the Great *Nest* of Being, to emphasize that each ascending level includes and transcends all lower levels. He depicts the Nest as a series of concentric circles. At the center lies matter, which is addressed by the discipline of physics; matter/physics is surrounded in turn by life/biology, mind/psychology, soul/theology, and finally spirit/mysticism. Wilber has called the Great Nest "a holarchy of extended love and compassionate embrace, reaching from dirt to divinity."

Wilber subdivides the Great Nest into four "quadrants" representing different aspects of reality: interior-individual (intentional), exterior-individual (behavioral), interior-collective (cultural), and exterior-collective (social). Each of these quadrants contains a dozen or so separate levels, some with esoteric labels such as *uroboric* and *typhonic.* With this scheme in place, Wilber can categorize those with alternate worldviews and demonstrate how far short they fall of full integral awareness. Scientific materialists who reduce all phenomena to mere matter are stuck way down in the *formop* (for formal/operational) stage, which Wilber derides as *Flatland.* New Agers who reject science and rationality in favor of superstitious worldviews are guilty of what he calls the *pre-trans fallacy,* in which primitive ignorance is conflated with spirituality; these pagans are relegated to the *magic* realm.

If you accept science and you also perceive the divine spirit dwelling within flowers, trees, clouds, and all of nature, you may have reached the *psychic* level, also known as nature mysticism. If you have a more abstract vision of God the creator, you have inched up to the *subtle* level, also called deity mysticism. *Causal* mystics have left the world of things behind and plunged into "the formless unmanifest," the void that lies beyond all things, even a personal God. At the very top of Wilber's hierarchy is "nondual awareness," in which the mystic returns to the world of things while remaining aware of the void that underlies all form.

Wilber warns that the path to enlightenment is riddled with pitfalls. As a result of an incomplete or poorly integrated mystical experience, you may think you are thoroughly enlightened when actually you are stuck in a lower level. You may succumb to depression, psychotic delusions, or narcissism. You may recoil in terror as you feel your ego dissolving into the mystical void, or you may become alienated from life instead of embracing it.

I have no doubt that if Wilber had been at the "Mystics" conference at the University of Chicago, he could have instantly categorized Teresa, Eckhart, and other mystics mentioned there according to his integral system. Teresa was stuck at the deity level, Eckhart at the causal level. Wilber could do the same with gurus of our era, including some of the holy fools discussed by Georg Feuerstein in his book *Holy Madness:* this one succumbed to narcissistic grandiosity, that one regressed to superstitious magic mysticism.

Wilber's prose is for the most part dense and analytical, laden with scholarly references; he does not wear his learning lightly. But now and then he steps out of his role as objective scholar and discusses mysticism in a more subjective, personal manner. In his 1999 book *One Taste,* which consists of excerpts from his personal journal, Wilber revealed that in 1995 he achieved nondual awareness, which persisted even while he slept. His awareness was occasionally disrupted by alcohol consumption or illness; it also dissipated, understandably, when he visited New York City to try to sell a book proposal. But otherwise, Wilber wrote, "I simply reverted to what I am, and it has been so, more or less, ever since."

To convey more vividly what cosmic consciousness feels like from the inside, Wilber sometimes veers into mystical lyricism. Some of these passages have an oddly chilly quality. "Yesterday I sat in a shopping mall for hours," he wrote in *One Taste,* "watching people pass by, and they were all as precious as green emeralds. The occasional joy in their voices, but more often the pain in their faces, the sadness in their eyes, the burdensome slowness of their paces — I registered none of that. I saw only the glory of green emeralds, and radiant Buddhas walking everywhere."

Reading such passages, I felt a queasy mixture of envy and disapproval. It struck me that seeing all humans as manifestations of some cosmic spirit is as reductionist as seeing them as clumps of quarks; it is the mystical equivalent of Flatland. It also struck me that my response was just sour grapes, and I pooh-poohed Wilber's mystical serenity because I knew it lay beyond my grasp. At any rate, the more I read by and about Wilber, the more eager I was to meet him. I hoped to get a clearer view of how things looked from his mystical mountaintop, and to resolve the ambivalence his writings aroused in me. I also wanted to get his reaction to the dark vision I had had in 1981, and to the peculiar theodicy I had derived from it.

Looking down on Flatland

Wilber lives in the mountains west of Boulder, Colorado, a college town northwest of Denver. He adheres to a daily regimen of physical, intellectual, and spiritual self-improvement. According to *One Taste*, he typically wakes up at four in the morning, meditates for an hour or two, reads and writes for eight or nine hours, and lifts weights for an hour. He rarely attends conferences, speaks in public, or grants interviews — not because he is reclusive (he told me later) but because he hates to be diverted from his work.

I thus felt fortunate when Wilber agreed to meet me at his home, which one journalist described as a cross between a Swiss chalet and a Japanese bathhouse. Driving my rental car from the Denver airport to Boulder, I was shocked at how much the landscape had changed since I lived in Colorado in the late 1970s. Back then, the highway connecting Denver and Boulder had passed through more than twenty miles of wide-open land, dotted only with cattle, cottonwood groves, and dilapidated ranches. Now the ranches had yielded to condo complexes and industrial parks. Denver's sprawl had fused seamlessly with Boulder's.

Most of Wilber's recent books feature photographs of him on their front covers. They show a rock-jawed, thin-lipped man, head shaved, peering sternly out from behind wire-rim glasses. Wilber was even more imposing in person. Over six feet tall, he was barefoot, wearing a sleeveless shirt that revealed his muscular torso. He was quite congenial, though. After a hearty handshake, Wilber escorted me into his cool, dimly lit house, which perches precariously on a treeless mountain slope overlooking Boulder. He had decorated his living room in refined bachelor style: black leather furniture, a huge bank of stereo equipment, an elevated platform heaped high with weightlifting equipment. Everything appeared clean, ordered, in its proper place. We sat down in a glass-enclosed porch that jutted disconcertingly out over the canyon. Looking east, I could see the mountains give way to the Great Plains. Flatland!

In person as in his writings, Wilber oscillated between pedantry and slangy informality, self-aggrandizement and self-deprecation. He mentioned that he had been a "very bright" and "all-American" youth, the valedictorian and president of his high school class and a star athlete. But he also called himself "geeky." When I asked if he had ever been tempted to become a guru, he replied, "I'm afraid if I did the guru thing, I'd be a victim of the Peter Principle. I would rise to the level of my incompetence and really start screwing some people up." He has a knack for writing, so he sticks with that.

Wilber asked whom else I had interviewed recently. He grimaced when I mentioned Steven Katz. Katz, Wilber complained, exemplifies the kind of postmodernism pervading academia. Wilber granted that postmodernism has made legitimate points about how social context contributes to the acceptance or rejection of certain ideas; in its extreme forms, however, postmodernism degenerates into "an endless play of narcissism and nihilism." Postmodernists claim that there is no truth, only "interpretations" imposed by "structures of power — so it's Eurocentric, logocentric, phallocentric." This claim undermines the authority not only of the perennial philosophy and other spiritual systems but of all approaches to reality, including science, Wilber said.

Unfortunately, many modern spiritual seekers share this hostile view of science, he said. They denigrate science or distort it to make it compatible with their preferred method of spirituality, whether psychedelic shamanism or eco-paganism or fundamentalist Christianity. Wilber understood the impulse behind this tendency. Some scientists assert that they can reduce all aspects of human psychology, including spirituality, to physiology or physics. "And that's when you get down to this absolutely meaningless, dumbass universe," Wilber said. This is the purely materialistic worldview that so upsets his mentor Huston Smith, and that Wilber calls Flatland.

But the proper response to scientific materialism, he said, is not to transform science into a pseudospiritual philosophy, as books such as *The Tao of Physics* try to do, or to call for a "new paradigm" of science that permits supernatural phenomena. One should simply accept that science addresses the material world and mysticism the world of spirit. You cannot understand the mystical realm by studying physics, psychology, philosophy, or any other intellectual discipline; you can gain entrance to that realm only by engaging in a spiritual practice that transforms your consciousness. Spirituality "is not just thinking about the world differently," Wilber said. "It's changing the *thinker.*"

Wilber warned me that there are no shortcuts to mystical transformation. He was less sanguine than Huston Smith about the spiritual benefits of entheogens. Wilber had tried marijuana a half-dozen times in his youth and LSD once; the experiences were not pleasant, let alone spiritually transforming. Drugs can trigger a "genuine breakthrough" in some people, Wilber said, but they cannot lead to stable, long-term spiritual growth. In fact, drug takers often become enmeshed in "increasingly bizarre fantasies about what spirituality is and isn't. It's *really* a problem." Meditation, Wilber believes, provides a much safer, more reliable route to mystical transformation than psychedelic drugs.

Wilber described spiritual practice in athletic terms. "It's like lifting weights and exercising muscles. The more you do it, the bigger the muscle gets." Early on in the practice of meditation, you generally experience only brief mystical flashes; making these states endure requires serious effort. One of the highest goals of mystical adepts is to remain in nondual awareness not just during meditation but continuously, even in the midst of deep sleep. This condition is akin to lucid dreams, in which you realize you are dreaming without waking up. After attaining this ability in 1995, Wilber sustained it until about a year ago, when a nasty staph infection left him bedridden for six months. "I lost a great deal of access to it," he said, but "it's slowly coming back."

Wilber made sure to disabuse me of some popular misconceptions about enlightenment. It doesn't instantly reveal all of the universe's secrets, as some mystical enthusiasts have implied; if you want to learn about physics and cosmology, you must study these subjects. More important, enlightenment doesn't instantly transform you into a saint or bodhisattva, a moral paragon dedicated to others' salvation. That lesson has become painfully obvious over the past few decades, as scores of supposedly enlightened gurus have been accused of immoral and even criminal behavior. Mystical maturity does not necessarily lead to psychological maturity, Wilber explained. In fact, powerful mystical experiences can retard psychological development by giving you delusions of grandeur.

"You get this experience: 'Oh my God! I am one with God! Oh, this is amazing! Nobody's *ever* had this experience in the entire history of the universe!'" But "the personality that you had before you got your satori is the personality you're stuck with. If you're a geeky little toad, then you're gonna be a geeky little toad that thinks he's God. And then it's going to be really hard to get rid of your geeky toadness," Wilber added, "because nobody can tell God what to do." There's nothing wrong with having a healthy ego; it helps you get things done in the world. But when gurus proclaim themselves to be perfect masters no longer bound by ordinary human ethics, "things start to get very, very ugly. It happens all the time. Not some of the time. *All* the time."

Wilber had learned this lesson rather painfully. In the early 1980s he fell under the spell of the self-proclaimed avatar Da Free John, aka Bubba Free John, Da Love Ananda, Adi Da, Avatar Adi, or just Da — originally Franklin Jones from Long Island. Like two other crazy-wisdom gurus, Rajneesh and Trungpa, Da Free John reputedly advocated alcohol use and sexual experimentation as catalysts for achieving liberation; to help male followers overcome their romantic attachments, he copulated with their girlfriends or wives. After disaffected devotees filed charges against him for physical and sexual abuse and false imprisonment in 1985, Da Free John and a group of disciples left the United States and settled in Fiji.

Wilber, who once called Da Free John "a Spiritual Master and religious genius of the ultimate degree," now stresses that even the most enlightened gurus "have feet of clay — all, no exceptions." Zen teachers have an excellent method of dealing with students who start comparing themselves to Buddha or God. "They take the stick and beat the crap out of you. And after five or ten years of that, you finally get over yourself." Even if you achieve cosmic consciousness, you must still work on your nonspiritual dimensions, by seeing a psychotherapist, doing public service, reading books with an intellectual content, exercising, eating properly.

"If you have a broken leg, you're still going to have a broken leg. If you get fired from your job, you're still going to be fired from your job. If your wife just walked out on you, she's still gonna walk out on your ass." Nondual awareness does not make your problems vanish, Wilber said, so much as it distances you from them. Emotions pass through your awareness "like clouds in the sky," he said, pointing at the roof of our glassed-in chamber. "You have a sense of skyness and not a sense of contracting on everything that comes along."

Enlightenment can also help you come to terms with your mortality. "To the extent that you stay relaxed into this open state, death doesn't have this overpowering terror." But nondual awareness does not resolve the mystery of death, as least not in Wilber's case. "There is a great Zen koan where a Zen master was asked what happens when we die. The Zen master says, 'I don't know.' And the student says, 'But you're a Zen master.' And he says, 'Yes, but not a dead one.'" Wilber chuckled. "I'm not a Zen master, but I can relate to that." Unlike "*really* enlightened people like Shirley MacLaine," he added drily, he had no memories of past lives. His "theoretical belief" is that each individual soul, or "small self," keeps reincarnating until it merges into the eternal, universal spirit. "Then, if it wants to reincarnate or not, it's literally a choice at that point."

Wilber's mystical experiences have given him a fundamentally optimistic outlook. He agrees with Aldous Huxley that reality, in spite of its dark, painful aspects, is on the deepest level All Right, "capital A, capital R." Wilber also believes that human history shows slow but steady material, intellectual, and spiritual progress. "It's sort of like the stock market," he explained. "You know that over the long haul there is a slow upward trend, even though there are great depressions and nightmares and all this kind of stuff. If you go back and look at the amount of suffering human beings have had over the last million years, in *so* many ways it's gotten better and better and better."

As Wilber spoke, I found myself reflexively agreeing with him: Yes, life is getting better and better as a result of science and medicine and advances in human rights. Then I remembered the ugly urban sprawl I had passed on my way from the Denver airport. All the other threats we face besides overpopulation and environmental degradation came crowding into my head: AIDS, racism, religious fundamentalism, weapons of mass destruction. As if reading my mind, Wilber cautioned that our evolution is fraught with hazards; continued progress is by no means guaranteed. "As society adds levels of depth, there are more things that can go wrong at every stage," he said, including "this whole planet going up in smoke."

Wilber's evolutionary, progressive view of human history is apparently embedded in a cyclical Hindu cosmology. From the highest mystical perspective, Wilber said, the history of humanity and even of the entire universe shrivels into insignificance. "Fifteen billion years is *nothing*. I mean, just the blink of an eye." He snapped his fingers. "The mystics are pretty sure about this. We've been doing this forever. This has happened a gazillion times; it'll happen another gazillion." According to Hindu myth, Brahman, the cosmic source of all things, creates a new universe with every exhalation. But Brahman itself — the spirit or mind from which all things spring — is eternal.

This, Wilber said, is the fundamental insight that mysticism gives us and that materialistic Flatlanders have such a hard time accepting: Mind is not just an ephemeral epiphenomenon of matter; it is eternal. What if we humans are the only conscious beings in the universe and an asteroid strikes the earth and wipes out us and all other life? I asked. Where will spirit be then? Won't the universe be reduced to Flatland, to dead, consciousless matter? Mystics are "pretty unanimous," Wilber replied, in believing that mind rather than matter is the basis of reality and hence can never be extinguished. The material universe "is a manifestation of this pure awareness" perceived in deep meditation.

THE BODHISATTVA SWATS A BUG

There is much to admire in Ken Wilber. He speaks openly about the limitations of mystical experience, acknowledging that even enlightenment does not lead to perpetual bliss and moral perfection. He can be candid, startlingly so, about his own shortcomings. In *Grace and Grit,* Wilber recounted his marriage to Treya Killam, who learned she had breast cancer one week after their marriage in 1983. (The marriage was Wilber's second; his first ended in divorce in 1981.) He confessed that at one point during Treya's illness he became suicidally depressed. He stopped meditating and writing and started drinking heavily. Once, consumed with self-pity and resentment toward his wife for interrupting his intellectual and spiritual growth, he struck her. With the help of psychotherapy, they saved their marriage, and they were more in love than ever when Treya died in 1988.

Wilber possesses a vigorous, wide-ranging mind, and unlike most spiritual authors he does not pander to his audience, although he could almost certainly sell more books if he dumbed them down or catered to New Age sentiments. Wilber's work has been favorably reviewed by *Skeptical Inquirer,* which normally eviscerates writers who give off even the faintest whiff of New Ageism. His critique of postmodernism — and in particular of the position of Steven Katz and other anti-perennialists — is cogent. He displays an impressive knowledge of and respect for science, real science, and he resists the temptation to transform it into a pseudoreligion. He excoriates the suggestion of some New Age authors that we can overcome any disease or hardship if our faith in our own minds is strong enough; this claim, Wilber points out, implies that it is our fault if we cannot cure our own cancer.

But something about Wilber irks me. During our meeting, I found myself looking for flaws even more obsessively than I ordinarily do with interviewees. At one point, a large winged beetle dropped on the table between us and scuttled about frantically. When Wilber swatted it away, I caught myself thinking: Aha! This bodhisattva is mean to bugs.

Perhaps it is Wilber's self-regard. For all his warnings about the perils of narcissism, or "psychic inflation," he does not seem to have entirely avoided that pitfall himself. Soon after I started reading his books, a sentence popped into my head, one that came back to me again and again during my research for this book: *I'm enlightened, and you're not.* Wilber keeps letting us know how highly evolved he is physically, psychologically, and intellectually, as well as spiritually. In *One Taste* he reprinted fawning letters from admirers. His self-deprecating asides seemed aimed only at making us admire his modesty.

In his books Wilber occasionally refers to himself as "I-I," which denotes both his puny, individual, mortal self and the cosmic, eternal Self, the "Seer that cannot itself be seen." This guy's ego is so big, a friend familiar with Wilber's I-I conceit once told me, that one I isn't enough for him. In person, Wilber was more modest and amiable than I had expected. But he did make it clear that vanishingly few people have reached his level of enlightenment. Not even the Dalai Lama can sustain nondual awareness through deep sleep, Wilber informed me, as he can.

Wilber also violates the perennialist concept that ultimate reality is mysterious, indescribable, wondrous. He does pay lip service to this principle. "If Spirit does exist," he declares, "it exists in the direction of wonder." He defines enlightenment as "not 'omniscience' but 'ascience,' not all-knowing but not-knowing." But these declarations do not ring quite true in context. Wilber speaks with an authoritarianism, pedantry, and didacticism that imply omniscience, not ascience; knowing, not unknowing. Confronted with Wilber's dogmatism, I wondered whether postmodernists such as Bernard McGinn and Steven Katz might have a better grasp of mysticism after all.

Wilber is well aware that he strikes some people as a know-it-all. At one point in our conversation, I remarked on the confidence with which he sets forth his opinions. I then added that he seems to be . . .

"An arrogant asshole!" he blurted out.

No, I said, startled, that wasn't where I was going.

"That's usually it," Wilber said with a rueful grin.

What I was going to say, I continued, was that Wilber seems to be implying that his version of the perennial philosophy might be a final framework for understanding reality.

"Oh, no," Wilber assured me, shaking his head. "Every moment transcends and includes the previous one. So even if every single thing I say is right, the next moment will transcend and include that. It unfolds forever."

But at the end of our conversation, Wilber's enormous self-assurance — his sense that he really has a grip on this reality business — emerged quite starkly. We had talked for almost five hours. The sun had slipped behind the mountains. I said I should probably be going; he had already been more than generous with his time. In fact, I was exhausted, physically and mentally. I wanted to drink a beer, watch a football game on television, anything that would distract me from pondering ultimate reality.

Somewhat to my dismay, Wilber said he wanted to make a couple more points. He sipped from a glass of water and paused a moment to collect his thoughts. Given my questions, he began, he suspected that I viewed mysticism as some strange, exotic phenomenon. Actually, mystical awareness is simply the culmination of human cognitive development. His latest book, called *Integral Psychology*, makes that clear.

Wilber fetched the book and flipped through a twenty-one-page table summarizing the worldviews of scores of mystics, philosophers, and scientists — from Plotinus to present-day developmental psychologists such as Robert Kegan, Lawrence Kohlberg, Carol Gilligan, and Howard Gardner — who have found universal patterns in our intellectual, moral, and spiritual growth.* This table shows that "there is a commonality, cross-culturally, up to various sophisticated modern researchers," which supports

---

*Howard Gardner has implicitly expressed doubts about Wilber's integral psychology. The Harvard psychologist is best known for proposing that we have not a single, all-purpose intelligence but multiple "intelligences" in domains such as language, mathematics, music, and interpersonal relations. In *Intelligence Reframed* (Basic Books, New York, 1999), he considered whether to add "spiritual intelligence" to his list. Certain individuals, he noted, seem particularly adept at achieving spiritually significant altered states through yoga, meditation, or other means; these same individuals can often discuss profound existential questions in a way that appeals to others. Gardner eventually decided not to add spiritual intelligence to his list of intelligences. He noted that as a secular Jew and scientist, he was "as much frightened as intrigued by people who see themselves (or who are seen by others) as spiritual individuals. I fear the strangeness of the beliefs to which they may subscribe, and I fear the effects that charismatic figures (in the style of Jim Jones or David Koresh) may exert on often hapless followers" (p. 65). Gardner was raising the troubling issue of crazy wisdom, the tendency of certain spiritual leaders to act in ways that seem perverse or evil rather than saintly.

Wilber's hierarchical model of reality. "It becomes one of the closest things we have to something like a periodic table of elements. Even though we can refine the understanding down the years, nobody is going to come along and go, 'Guess what, silver doesn't exist.' It's just not going to happen." Wilber thinks he has the final framework after all: a mystical periodic table.

Solving the Nirvana Paradox

Wilber's hierarchical — or *holarchical,* to use the term he prefers — model of human development is a more sophisticated version of the one laid out by the Canadian psychiatrist Richard Bucke a century ago in *Cosmic Consciousness.* Enlisting the relatively new theory of evolution in his mystical cause, Bucke proposed that our distant ancestors perceived the world directly, with little more reflection or self-awareness than animals. As humanity evolved, this "simple consciousness" was superseded by a more sophisticated "self-consciousness." In the same way, Bucke proclaimed, self-consciousness will eventually yield to cosmic consciousness, and the entire human race will become enlightened.*

Bucke granted that only a few extraordinary individuals had thus far attained cosmic consciousness. These were "mostly of the male sex, who are otherwise highly developed — men of good intellect, of high moral qualities, of superior physique."† They also tended to be white, or "Aryan." Initially, inferior portions of humanity — notably females and "the negro race" — would lag behind others in this spiritual evolution; the mystical psychiatrist also warned that expanded consciousness brings an increased

*Other perennialists have resisted placing mysticism within an evolutionary, progressive framework. In an essay introducing a translation of the *Bhagavad Gita* (New American Library, New York, 1951, pp. 16–17), Aldous Huxley deplored the West's growing infatuation with progress: "Attention and allegiance came to be paid, not to Eternity, but to the Utopian future." In *Forgotten Truth* (HarperSanFrancisco, 1992), Huston Smith was harshly critical of the attempts of the Jesuit anthropologist Teilhard de Chardin to fuse evolutionary theory and spirituality.

†A female author recently pointed out the obvious: "Apparently it did not occur to [Bucke] that men, enjoying in most cultures the advantages of literacy and power, may have recorded their ecstasies more often than women, or that women — who for centuries were burned as witches for what often earned men reverence — may have hesitated to report their experiences" (*The Ecstatic Journey,* by Sophie Burnham, Ballantine Books, New York, 1997, p. 84).

risk of insanity. But he had no doubt that in the long run evolution would result in a new race of powerful mystical beings: "This new race is in the act of being born from us, and in the near future it will occupy and possess the earth."

Wilber expresses similar sentiments, shorn of the sexist, racist, eugenicist, social Darwinist overtones. But while he agrees with Bucke that humanity is making steady progress — things are getting better! — Wilber acknowledges that this process is fraught with danger and could be interrupted at any time; our world could be destroyed by nuclear weapons or an errant asteroid. Ultimately spirit will endure in the universe, he says, but not necessarily the *human* spirit. The only real consolation Wilber offers is that of mystical awareness itself, which he makes the endpoint of human psychological development and indeed all of creation.

But Wilber has hinted that even enlightenment might not be totally consoling. That, at any rate, was how I interpreted a remarkable passage in his book *One Taste.* Wilber recalled that during one of his first meditative forays into the mystical void, he had "the vague feeling — very subtle, very faint — that I didn't want to be alone in this wonderful expanse." He speculated that a similar desire for companionship may be what motivated God to create the world, to divide His perfect oneness into the imperfect many. Wilber quoted another spiritual teacher who put it this way: "It's no fun eating dinner alone."

"Here you are," Wilber continued, "the One and the Only, the Alone and the Infinite. What are you going to do next? You bathe in your own glory for all eternity, you bask in your own delight for ages upon ages, and then what? Sooner or later you might decide that it would be fun — just *fun* — to pretend that you were not you. I mean, what else are you going to do? What else *can* you do?"

Elaborating on this idea, Wilber noted that as a child he liked playing checkers, so much so that he occasionally played against himself. The catch was that when he was playing red he had somehow to forget that he was also playing black, and vice versa. In the same way, Wilber proposed, God makes the game of creation interesting by manifesting Himself in individual minds that have "forgotten" their true source. This divine self-forgetting is sometimes called *amnesis* by Western theologians.

When I originally read this passage, I underlined it, circled it, put stars by it, and wrote "VERY IMPORTANT" in the margin. I even marked the page with a Post-it on which I wrote, "Why God creates." Wilber was addressing what I liked to call the nirvana paradox: If nirvana is so great, why does God create? The answer that I — and, I suspected, Wilber — had arrived at is that nirvana isn't so great. Even God gets lonely.

During my conversation with Wilber, I recalled his remarks about God not wanting to eat dinner alone and said I had a similar explanation of creation. I suspected that God creates not just for companionship or "fun" but because of His terrified recognition of His own solitude and improbability and even His potential death; God "forgets" Himself and flees into multiplicity because He cannot bear to confront His plight. I added that all this had come to me during a drug trip about twenty years ago, when I actually *became* this lonely, fearful God.

"That can't have been pleasant," Wilber said. During my trip I may have glimpsed a partial truth, he told me. The mystical void can indeed terrify you when you are on the outside looking in, because you fear the death of your limited, mortal ego. But once you are immersed in the abyss, "it's not terror at all. It's absolute, infinite liberation. It's *exquisite* in that sense. So no one has to worry about falling into that." On the other hand, "it's no fun just to be in that unmanifest state forever," Wilber said, so the cosmic spirit manifests itself as matter and mind and all the wondrous diversity of creation.

But why, I asked, would this fun-loving spirit create such a fucked-up world? If creation is motivated by God's terror and desperation, it might explain why existence contains so much pain and evil. "Okay," Wilber said, "you think this is because God is, like, a confused, tortured soul that spit all of this out?" God creates not only out of terror, I responded, but also for the pure joy of it, or for fun, as Wilber put it; creation's joyous and terrible aspects reflect this dual divine motivation.

The gnostics espoused a theology like mine, Wilber informed me. The gnostics looked at all the evil in the world and asked, What kind of God would create this? They postulated that the highest level of being is the Godhead, which is pure and good. But below the Godhead is the Demiurge, a "semi-evil, semi-divine" being who created the flawed world in which we live. "And the Demiurge was a confused boy, basically," Wilber said. "Tortured, semi-psychotic *nutcase*."

The gnostics sought to escape this flawed realm through gnosis, a "satori-type" awareness that vaulted them "beyond the Demiurge and the manifest world to pure Godhead." The trouble with gnosticism, Wilber continued, is that it encourages alienation from "the manifest world" — that is, the world in which we humans live. True spirituality embraces the manifest world instead of scorning it. Ultimate reality "is not a green monster thinking evil things," Wilber assured me with a smile. "It's *nothing.* There is literally nothing going on, and that's the Godhead. So it's not good and it's not bad." My dark, quasi-gnostic, drug-induced vision "at best was an intermediate-level phenomenon. You didn't see some deep, hidden side of God that we're going to get stuck with for eternity." He laughed heartily.

I wasn't entirely convinced by Wilber's response. He seemed to be denying the implications of his own experience, the loneliness he had felt when encountering the void. Moreover, gnostics were hardly alone in suggesting that creation's imperfections reflect some imperfection in God. This idea is implicit in the Kabalist concept of *tsimtsum,* which equates creation with God's self-alienation and fragmentation.* A similar notion has been advanced by Da Free John. Although he now sees Da Free John as a deeply flawed individual, Wilber still thinks the guru is a brilliant mystical philosopher. Da Free John once offered what seemed to be a gnostic explanation for his bizarre and hurtful behavior. "God is crazy just like me," he said. "In fact, God is crazy just like *you.*"

Meditation, Telescopes, and Brain Scans

On the other hand, Wilber's version of ultimate reality is certainly more hopeful, optimistic, and world-embracing than mine, and it appeals to common sense. The fear I felt during my drug trip was just fear of my self-dissolution and death, which in my grandiosity I projected onto God. If I were more mystically advanced, surely I would have embraced rather than fled from the void, and known the bliss and peace that Ken Wilber and many other mystics have attained.

Wilber's perennial philosophy still poses a serious practical problem, however. Nondual awareness may be perfectly consoling, as Wilber insists,

---

*Wilber's suggestion that God forgets Himself so that He can play with Himself (so to speak) is echoed by the Kabalist scholar Arthur Green of Brandeis University in his book *Seek My Face, Speak My Name* (Jason Aronson, Northvale, New Jersey, 1994). God divides Himself, Green suggested on p. 65, so that He can play a game of cosmic hide-and-seek, which "just doesn't work as solitaire!"

but only a small number of people ever achieve this supreme mystical state. Wilber has asserted that "radical transformative spirituality is extremely rare." He cited the estimate of two Zen masters that only about a dozen living Zen monks have reached the highest level of mystical knowledge, and only a thousand or so have achieved this acme throughout history.

At the same time, Wilber often implies that attaining mystical truth and consolation is relatively straightforward, a matter of finding the right spiritual discipline and sticking with it. During our conversation, he compared meditation to a scientific instrument. "If you want to see a cell nucleus, look down this microscope. If you want to see the moons of Jupiter, look through a telescope," he said. "On the spiritual side, if you want to see your Buddha nature, if you want to see Christ consciousness, if you want to see the religious side of the equation, fold your legs, sit down each day for two hours, count your breath from one to ten. Do that for five years and get back to me."

This comparison of meditation to a scientific instrument is a cornerstone of Wilber's argument — laid out most explicitly in his book *Marriage of Sense and Soul* — that mysticism is in its own way as empirical as science. I told Wilber that I didn't buy the analogy. Although everyone who looks through a telescope can see the moons of Jupiter, not everyone who meditates sees the same mystical reality; you can see heaven, hell, visions, or nothing at all. Wilber's mentor Huston Smith, in all his decades of meditation, never had an experience as profound as those induced by psychedelic drugs, and Wilber himself meditated for more than twenty years before achieving permanent nondual awareness.

Wilber granted that meditation is not reliable at inducing mystical states. Moreover, reaching higher levels becomes progressively more difficult, and the risk of succumbing to a spiritual pathology such as narcissism or gnostic alienation becomes progressively greater. Full-blown enlightenment, he said, may always be reserved for an elite few, just as the esoteric truths of particle physics are. But he foresaw a day when mystical equanimity might become more accessible as a result of investigations into the neurophysiology of mystical states.

Wilber was the subject of an experiment involving an electroencephalograph, which records electromagnetic oscillations in the brain detected by electrodes placed on the scalp. The EEG readings showed Wilber's brain waves shifting as he accessed different levels of consciousness. The videotape he made of this experiment has convinced "virtually every scientific type I show it to," Wilber said, that "primordial awareness is not just an idea you memorize but the result of actual practice that truly changes your very makeup."

Neuroscience might help skeptics and believers alike gain firsthand knowledge of the mystical realm, Wilber said. Electroencephalography provides a way to measure the spiritual progress of people following various spiritual practices, such as Zen, kundalini yoga, or Vipassana. Eventually, such studies might "dramatically increase people's actual chance of interior transformation" by matching individuals to the mystical techniques most suited to their particular needs and constitutions. As scientists learn more about the neurophysiology of mystical experiences, moreover, they might be able to design more efficient mystical technologies.

In recent years brain scientists have acquired instruments more powerful than the electroencephalograph, which is a relatively crude, low-resolution technology that can detect only large-scale electromagnetic fluctuations at the surface of the brain. Researchers are employing magnetic resonance imaging, positron emission tomography, and other methods that can detect neural activity deep in the brain with much higher resolution. A growing number of scientists — sometimes called neurotheologians — are now attempting to unravel the brain processes underlying mystical experiences. Far from trying to explain away mysticism, they hope to make its claim to truth even more compelling.

4
Can Neurotheology Save Us?

On the May 7, 2001, cover of *Newsweek,* a full-lipped female model gazed raptly upward with turquoise, mascara-lined eyes. She held her hands skyward, too, as if to cup the brilliant white light descending upon her. The headline over this image stated: "God and the Brain: How We're Wired for Spirituality." The story within reported on scientists' efforts to pinpoint the regions of the brain underlying religious and mystical experiences.

"Although the field is brand new and the answers only tentative," the article announced, "one thing is clear": There is a "common core" of spiritual experiences that is "consistent across culture, across time and across faiths." That is precisely what perennialists — Aldous Huxley, Huston Smith, Ken Wilber — have been saying all along, and what anti-perennialists such as Steven Katz have doubted.

The researcher featured most prominently in the *Newsweek* article was Andrew Newberg, a young radiologist at the University of Pennsylvania Medical School. Newberg's mystical quest began in the early 1990s when he met Eugene D'Aquili, a professor of psychiatry at the same university who had been studying the physiological underpinnings of mystical experiences since the 1970s. Newberg, then a medical student specializing in radiology and nuclear medicine, shared D'Aquili's interest in mysticism and offered to help him test his theories with sophisticated brain-scanning technologies.

The two began a collaboration that yielded several papers and two books before D'Aquili died of a heart attack in 1998. Some of their suppositions, stated baldly, make Newberg and D'Aquili sound like those whom Ken Wilber derides as Flatlanders, materialistic killjoys intent on reducing mystical states to mere neural eructations. They traced sensations of unity to suppressed activity in a particular part of the brain, and they discerned a biological resemblance between mystical ecstasy and orgasms (which made the erotic tinge of the *Newsweek* cover all the more appropriate). And yet Newberg and D'Aquili proposed that their work was lending credence to mystical claims. Like Wilber, they were proponents of the perennial philosophy, although they never used that phrase in their writings. They preferred to call their approach neurotheology, a term they helped to popularize.*

The common element in all spiritual experiences, Newberg and D'Aquili contended, is a sense of unity deeper than that conveyed by ordinary consciousness. This sensation can range from the mild communion that a congregation feels while singing a hymn to the "state of absolute unitary being" in which you lose all sense of self, of subject-object duality. This state (which Ken Wilber has called causal awareness and others have named the introvertive mystical experience) "seems to be identical in all places and times of which we have record," Newberg and D'Aquili wrote, echoing the rhetoric of the perennial philosophy.

Newberg and D'Aquili divided all methods for achieving unitive experiences into two categories: top-down and bottom-up. Top-down methods, which include meditation and prayer, achieve transcendence through relaxation, by focusing and calming the mind. Bottom-up techniques, which include dancing, hyperventilation, chanting, and vigorous yoga, approach the same goal through excitation. Each method taps into a different component of the body's autonomic nervous system, which regulates heartbeat, blood pressure, respiration, metabolism, and other physiological functions. Top-down methods exploit the so-called quiescent component of the autonomic system, which limits the body's expenditure of energy and maintains its equilibrium. Bottom-up methods exploit the arousal component of the autonomic system, which triggers the body's fight-or-flight response, causing adrenaline to be pumped into the bloodstream, boosting heart rate and respiration.

If either the arousal or the quiescent component is pushed far enough, the one activates the other through a "spillover effect," producing

---

*In *The Mystical Mind* (Fortress Press, Minneapolis, 1999, p. 4), Newberg and D'Aquili credited the scholar James Ashbrook with introducing the term *neurotheology* in 1984. In fact, Aldous Huxley used it more than twenty years earlier in his utopian novel *Island* (Harper & Row, New York, 1962, p. 144).

a paradoxical state of ecstatic serenity. At the same time, activity decreases in a region at the top and rear of the brain called the posterior superior parietal lobe. Newberg and D'Aquili referred to this region as the orientation-association area, since it helps us orient our bodies in relation to the external world. Patients whose posterior superior parietal lobe has been damaged often lose the ability to navigate, because they have difficulty determining where their physical selves end and the external world begins. Suppressed activity in this region, Newberg and D'Aquili hypothesized, could lead to the decreased sense of subject-object duality and the heightened sense of unity with the external world reported by mystics.

The two scientists tested their hypothesis by scanning the brains of eight American Buddhists practicing a form of Tibetan meditation and three Franciscan nuns engaged in contemplative prayer. Most of the Buddhists and nuns displayed increased neural activity in the prefrontal cortex and decreased activity in the posterior superior parietal lobe, as Newberg and D'Aquili had predicted. They declared that as a result of their imaging studies, skeptics "can no longer dismiss the description of [mystical] states in the world's religious and mystical literature as 'the silly imaginings of religious nuts.' They must be accounted for and their claims and practical implications carefully examined." Their brain scans prove that the accounts of mystics are based "not on delusional *ideas*, but on experiences that are neurologically real," Newberg and D'Aquili stated in their book *Why God Won't Go Away.*

Newberg and D'Aquili did not say flatly that they had proved the truth of unitary visions — and of God — but they came close. I once heard an astrophysicist compare his satellite images of the so-called cosmic background radiation — microwaves that suffuse the universe and are thought to be the faint afterglow of the big bang — to a mystical experience. "If you're religious," the astrophysicist said, "it's like looking at God." Newberg and D'Aquili found evidence of divinity not in deep space but in the human brain. When they proposed in *Why God Won't Go Away* that their brain-imaging work has yielded a "photograph of God," they intended no irony.

In their book *The Mystical Mind,* Newberg and D'Aquili made an extraordinarily bold prophecy: "It is possible that with the advent of improved technologies for studying the brain, mystical experiences may finally be clearly differentiated from any type of psychopathology." By looking at brain-scanning images, in other words, neurotheologians might distinguish genuine mystical visions from pathological gnostic delusions, crazy wisdom from mere craziness. Perhaps they will even resolve the debate between perennialists and postmodernists over which mystical visions are transcendent revelations and which are products of the mystic's cultural indoctrination. Age-old disputes over mysticism's meaning might finally be settled by an image showing decreased metabolic activity in the posterior superior parietal lobe.

Andrew Newberg's doubts

I met Newberg at the Hospital of the University of Pennsylvania on a bone-chilling, snowy day. I was feeling cantankerous — even more than usual, that is. I had spent the previous evening visiting a friend I'd known since high school, whom I'll call Pete. For several months, he had been undergoing chemotherapy for non-Hodgkin's lymphoma. Pete was scheduled to receive a bone-marrow transplant. His doctors said it was his best chance for survival, but it could also kill him. Reviewing Newberg's research just before the interview, I had wondered what possible consolation Pete — and his wife and two young children — could get from neurotheology as they faced the possibility that he might soon die.

As the interview unfolded, however, I found myself disarmed by Newberg's amiability and earnestness — and his modesty about his own research. We sat in a conference room in the radiology department lined with medical texts and journals and cluttered with equipment: computer monitors, viewgraph machines, slide projectors. White-coated colleagues occasionally whisked through the room. This was an unlikely setting for a conversation about God, ultimate reality, and states of absolute unitary being.

Newberg also seemed awfully young and, well, too normal to be delving into such esoteric affairs. With his curly brown hair and eager-to-please demeanor, he looked much younger than someone in his mid-thirties. He wore a white lab coat, blue oxford shirt, sensible brown shoes, wire-rim glasses, ID card, and a pager clipped to his belt. Newberg is aware of the impression he conveys. The reason he did not become a psychiatrist like his late partner D'Aquili, Newberg joked, is that he "wasn't crazy enough."

Newberg traced his interest in philosophical matters to his childhood. His parents, Reform Jews, taught him that faith is important, but they also encouraged him "to ask a lot of questions." As a youth he studied

Western and Eastern philosophy as well as science. He developed the habit of lying on his back with his eyes closed and pondering the questions his readings raised. "It started out with a Western scientific approach to saying, What can you know without making false assumptions?" His questioning gradually turned into "a purely introspective thing," in which he was "deconstructing the universe" and placing "all these different things into doubt." Newberg was reluctant to elaborate on this practice. "I don't like to describe it as a meditation per se," he told me. "It's just my own approach to asking questions, I guess."

I asked Newberg if he had ever had a mystical experience. "Nothing I could really say was totally unusual," he replied; certainly no states of absolute unitary being. Beyond feeling that there is "some sort of connectedness to the universe," he had no specific spiritual beliefs, or at least none he wanted to disclose. He admitted that he worries his research will be dismissed as biased if he reveals too much about his personal spirituality.

Newberg nonetheless acknowledged his hope that contemplative modes of investigation can complement scientific ones. "I am interested in seeing where science goes, seeing where its limitations are, and then seeing what you can do after reaching those limitations." When strictly rational, empirical analysis reaches an impasse, he said, meditation and other non-intellectual types of cognition might indicate new avenues for exploration.

Newberg already saw points of convergence between mystical and scientific approaches. For example, he said, quantum mechanics suggests that even an absolute vacuum brims with energy, which constantly generates "virtual particles" that exist for an instant before vanishing; some cosmologists have speculated that the universe might have begun as a virtual particle generated by a primordial vacuum energy. Buddhists have described the relationship between the void and the world of forms in similar terms, Newberg said.

Quantum mechanics also provides a way to reconcile hard-core scientific materialism with mystics' conviction that mind rather than matter is primary. Both a purely materialistic view of reality and a purely mind-based view pose problems, Newberg said. "If you've got matter only, how does consciousness arise? If you've got consciousness only, how does matter arise out of that? I think to a certain extent there may be a way to integrate those two concepts."

The paradoxical finding of quantum mechanics about the nature of light "provides a nice analogy," Newberg said. Light appears to be either particles or waves, depending on how you view it. In the same way, a mystical experience can be seen either as an excitation of nerve cells or as a strictly mental phenomenon, depending on whether you look at a brain scan or listen to the mystic describe her experience. The fact is, Newberg said, such an experience is both physical and mental.

In defending the validity of mystical insights, Newberg reiterated arguments that he and D'Aquili made in their books. Although mystical experiences are quite diverse, most spiritual traditions agree that the most profound experience of all is the state of absolute unitary being. "Once you come out of that," Newberg said, "your way of describing it and understanding it may be very culturally based or religiously based. But the actual experience of that most extreme state *has* to be the same, because it's total unity, so there's nothing else it can be."

This is essentially the same position set forth by Huston Smith and Ken Wilber. Newberg speculated that other mystical experiences, negative or positive, occur as people "are heading down that continuum" toward the state of absolute unitary being "but not getting all the way." (Newberg differs from Wilber in one important respect: Newberg suggests that absolute unitary being is the supreme mystical condition, whereas Wilber asserts that there is an even higher state, nondual awareness, in which perception of the void underlying all things is integrated with worldly perceptions.)

Another reason to take mystical experiences seriously, Newberg said, is that from a subjective perspective they feel truer, more *real,* than dreams, hallucinations, even ordinary perception; they seem to represent "a more fundamental reality than the baseline reality." The mystic's conviction that what she has seen is real persists after the mystical vision passes. Moreover, these subjective states correlate with observable brain states, as Newberg's brain scans of meditators and nuns have shown.

Mystical experiences are also good for us, Newberg said. In *Why God Won't Go Away,* he and D'Aquili cited studies associating mystical experiences with "higher-than-average levels of overall psychological health, expressed in terms of better interpersonal relationships, higher self-esteem, lower levels of anxiety, clearer self-identity, an increased concern for others, and a more positive overall outlook on life." Newberg and D'Aquili suggested that natural selection may have embedded a propensity for mystical experiences in our ancestors because of these positive benefits.

In short, Newberg and D'Aquili offered four reasons for believing mystical claims: mystical experiences occur in identical forms in all cultures, they feel truer than any other experiences, they have neurological correlates, and they are natural and beneficial. I had come to my interview with Newberg prepared to challenge his reasoning. After all, schizophrenic psychosis meets three of the criteria. Schizophrenia is a cross-cultural, biologically based phenomenon; schizophrenics are often certain that their delusions are true; and their delusions can be traced to specific brain states. But few rational folk would claim on this basis that schizophrenics' visions are true.

As for the assertion that mystical visions are good for us, most of the evidence cited by Newberg and D'Aquili involves the benefits of meditation or religious belief rather than mystical experiences per se. As the history of crazy-wisdom adepts shows, mystics may behave in ways that are harmful to themselves and others. Even if natural selection did embed a capacity for mystical visions in us because of their adaptive value, does that mean we should accept the validity of mystical claims, as Newberg and D'Aquili suggested? Or accept the persistence of religion, which supposedly sprang from our innate mysticism? According to that reasoning, we should also accept racism, xenophobia, and violent aggression, since our propensity for such behavior arguably has an evolutionary basis.

To my surprise, Newberg either anticipated all these points himself or conceded them when I raised them. He acknowledged, for example, that mystical visions are not always beneficial. Mystical and religious experiences may lead not to wisdom and compassion but to narcissism, fanaticism, or hatred. If you identify the object of your mystical union as Jesus or Muhammad, Newberg said, you may despise or want to kill someone who does not share your passion for those religious figures.

Moreover, self-dissolution may be experienced in a positive way, as a unification with something larger, or in a negative way, as fragmentation, as "the actual death of the ego," Newberg said, "which is a horrible, scary thing." Although popular books on near-death experiences make them sound universally heavenly, several studies have shown that as many as half of them are hellish. Some people find themselves trapped in a "terrifying emptiness."

His former colleague D'Aquili once told Newberg about a sect in the Far East that deliberately sought negative states of absolute unitary being, in which ultimate reality is perceived as totally evil rather than totally good. "You're almost talking about the devil," Newberg said. My ears pricked up. D'Aquili seemed to have been referring to something akin to gnosticism, in which reality is perceived as All Wrong rather than All Right. Where was that sect? I asked. "I never got the name of it," Newberg said ruefully. If D'Aquili had had that information, he took it to the grave with him.

Inspecting nuns' brains

Newberg did not oversell his brain-imaging research. He and D'Aquili scanned the brains of the meditators and nuns with a technique called single photon emission computed tomography, or SPECT, a variant of the better-known positron emission tomography, or PET. The chief advantage of SPECT is that it can capture the brain activity of meditators in a relatively natural setting. The subject meditates not in the SPECT chamber itself but in another room. When a nun, say, reaches the peak of her contemplation, she pulls a string tied to her finger. At that signal, a fluid tagged with a short-lived radioactive isotope flows into her through an intravenous tube. The fluid quickly travels to her brain and is absorbed by its most active regions. The nun is then taken to the SPECT chamber and scanned by a robotically controlled camera. The resulting image reveals levels of neural activity not during the scanning session but immediately after she was injected with the radioactive fluid, when she was presumably still immersed in contemplation.

Newberg granted that the experimental setting is quite artificial. Some subjects find the intravenous needles uncomfortable, and they have to remain self-conscious enough to remember to pull the string when they reach what is supposed to be their deepest state of awareness. Another disadvantage of the SPECT method is that it provides only a single snapshot of the brain; it cannot reveal all the complex changes that take place during a session of meditation or prayer.

"With one point in time, you may or may not catch what you're looking for," Newberg said. Ideally, he would like to observe how the brain changes throughout the meditative process. But that would require keeping the subjects in the scanning device, which might distract them from their meditation. A more generic limitation of his experiments is that he must rely on his subjects' assessments of their own states of mind; one subject's self-described peak experience might be a trivial event for someone else. "Try as hard as you might, there's no clear way of measuring" a subjective experience, Newberg said.

He conceded that his study might be capturing brain states related only to meditation and prayer — and only two specific types at that, out of the many that are practiced — rather than to deep mystical states per se. Many people have their deepest spiritual experiences when they are not engaged in any spiritual practice; they may be making love or walking in the woods or listening to music. Newberg nonetheless hoped that continued research by himself and others will reveal that similar brain structures and processes underlie mystical experiences induced by different means. "We chose meditation because it's reproducible," he said. "We can bring a person into a laboratory and say, 'Okay, start meditating now.' There is no way to really measure spontaneous mystical experiences, unless you put electrodes on everyone walking around and see who has one."

The more Newberg talked about his neurotheological research, the more I wondered what good he thought it could do. In *The Mystical Mind,* he and D'Aquili expressed hope that refinements in brain-scanning technology might allow "mystical experiences to be differentiated from any type of psychopathology." Newberg initially seemed prepared to defend this possibility during our conversation. He cited a report that brain scans of schizophrenics vary depending on whether they are hearing a real human voice or an unreal, hallucinatory voice. This experiment, Newberg said, demonstrates that neuroimaging can distinguish real perceptions from delusional ones.

But hallucinations and dreams, I responded, can impart meaningful and true information. One famous example involves the nineteenth-century German chemist Friedrich Kekulé, whose dream of a snake swallowing its own tail inspired him to guess the correct structure of the benzene molecule.* Newberg nodded in agreement. He then essentially retracted his and D'Aquili's earlier prediction that one day neuroimaging would distinguish "true" mystical visions from delusions. "I don't think *any* brain-imaging study could do that," he said.

If a scan reveals a tumor in the brain of a nun having religious vi-

---

*A recent article, "What Dreams May Come?" by Paolo Mazzarello (*Nature,* November 30, 2000, p. 523), described Kekulé's dream and two other cases in which dreams inspired scientific breakthroughs. The Russian chemist Dmitry Mendeleyev traced his intuition of the periodic table to a dream in 1869: "I saw in a dream a table where all the elements fell into place as required. Awakening, I immediately wrote it down on a piece of paper." Similarly, the physiologist Otto Loewi said that during a night of fitful sleep in 1921, he envisioned the experiments on the chemical transmission of nerve impulses that later won him the Nobel Prize.

sions, Newberg said, that does not necessarily invalidate her visions; after all, there is evidence that some of history's greatest religious figures suffered from epilepsy, schizophrenia, and other disorders. Muhammad's visions of Allah might have been induced by seizures, Newberg said, yet he created one of the world's great religions. In other words, we cannot judge a vision's validity solely by its neurological basis; each vision must be weighed on its own merits.

Newberg was reiterating a position spelled out by William James a century earlier in *The Varieties of Religious Experience.* In a chapter titled "Religion and Neurology," James concluded that neurology can neither prove nor disprove the validity of specific mystical visions. If a mystic suffers from mental illness or brain damage, that doesn't mean that her visions are worthless delusions, any more than the mental illness of poets, artists, or scientists invalidates their achievements. We must judge these visions by their philosophical and moral consequences. "By their fruits ye shall know them," James added, "not by their roots."*

Another possible application of neuroscientific studies of mystical states would be finding more efficient methods for inducing them; Ken Wilber mentioned this possibility during our interview. In *Why God Won't Go Away,* Newberg and D'Aquili occasionally implied that we should have no need for new and improved mystical technologies. They declared that our brains are "mystical by default" and "wired" to experience God; that is why most people around the world, and more than 90 percent of Americans, profess belief in God.† But their own figures demonstrated that our belief in God is much greater than our propensity for mystical states.

Newberg and D'Aquili cited a 1975 survey by the Catholic priest and

---

*James led up to this quote in *Varieties* (Macmillan, New York, 1961, p. 34) with a discussion of Teresa: "Saint Teresa might have had the nervous system of the placidest cow, and it would not save her theology, if the trial of the theology by these other tests should show it to be contemptible. And conversely if her theology can stand these other tests, it will make no difference how hysterical or nervously off her balance Saint Teresa might have been when she was with us here below" (p. 33). Later (p. 276), James was surprisingly critical of Teresa, complaining that her professions of selflessness and humility obviously concealed a powerful self-regard and ambition, and that "her idea of religion seems to have been that of an endless amatory flirtation" between the devotee and God.

†Ninety-six percent of adult Americans believe in God, 90 percent believe in heaven, 79 percent believe in miracles, and 72 percent believe in angels, according to *Why People Believe Weird Things,* by Michael Shermer (W. H. Freeman, New York, 1997, p. 275). (Shermer, a histo-

sociologist Andrew Greeley, who found that roughly one third of adult Americans have felt "very close to a spiritual force that seemed to lift you out of yourself" at least once in their lives. Nevertheless, the vague wording of Greeley's question makes it a dubious measure of true mystical experience. Even if you accept that figure, it still means almost two thirds of the population has never had a single mystical experience. These nonmystics include Greeley, who once told me, "I'm the least mystical person I know."

The state of absolute unitary being is very rare, according to Newberg. He himself has never attained it, nor, to the best of his knowledge, did D'Aquili — nor any of the Tibetan Buddhists and Catholic nuns studied in the SPECT experiments. A British survey carried out in the 1970s found that "experiencing in an extraordinary way that all things are 'One'" is the least frequent of all spiritual epiphanies.

I asked Newberg if he hoped his research might help more people gain entrance to the mystical realm by leading to improved mystical technologies. He shook his head. While "theoretically possible," he replied, that is not his goal. He noted that we already have time-tested spiritual practices in the form of meditation, yoga, chanting, and prayer, which have been refined over many centuries. Science may never discover a single, sure-fire method for inducing mystical experiences, because "we are all individuals, and we all respond to things differently," Newberg added. "I think there will never be a technology that will work for everyone."

This was another remarkable admission. If the variability of human brains and psyches precludes finding a mystical technology that works for

---

rian of science and a former devout Christian turned agnostic, is the director of the Skeptic Society and publisher of *Skeptic* magazine.) But other surveys undermine the notion that religious belief is "hard-wired." In "What Americans Really Believe" (*Free Inquiry*, vol. 19, no. 3, 1999, pp. 38–42), the political scientist George Bishop presented data showing that religious beliefs "vary considerably among developed nations." He concluded that "there is probably no biological propensity per se to hold such beliefs" (p. 39). Moreover, two polls reported recently in *Nature* (April 3, 1997, pp. 435–436; and July 23, 1998, p. 313) found that only 40 percent of a random sampling of American scientists believe in God; only 7 percent of the elite scientists belonging to the National Academy of Sciences believe in God. Those who argue that religious belief has a biological basis often cite research on identical twins by a group at the University of Minnesota, which has reported that genes account for about 50 percent of the variation in religiosity. For a critique of this twin research, see my book *The Undiscovered Mind* (Free Press, New York, 1999, pp. 141–146).

everyone, then it could also preclude finding a single neural mechanism that explains everyone's mystical experiences. At the very least, Newberg's admission raises doubts about his and D'Aquili's claim that brain scans of eight men and three women engaged in two spiritual practices have revealed the "biological basis for mysticism."

A recent article in the *Journal of Consciousness Studies* put Newberg and D'Aquili's brain-scanning experiments in badly needed perspective. In "Meditation Meets Behavioral Medicine," Jensine Andreson, a professor of theology at Boston University, reviewed more than five hundred papers and books on meditation published over the last half century, including the work of Newberg and D'Aquili. Andreson cautioned that there are literally thousands of techniques that could be categorized as meditation; it is virtually impossible to define the term in a way that does justice to this vast diversity.

Not surprisingly, she said, attempts to measure meditation's neurological effects with brain wave monitors, positron emission tomography, and other techniques have yielded widely divergent findings. Meditation has been "prodded and poked by a variety of technological apparati, with inconclusive results," Andreson commented. For every reported finding there is a conflicting result. In a study of meditators by Danish researchers, PET scans showed increased activity in the parietal lobes and decreased activity in the frontal lobes — the opposite of what Newberg and D'Aquili saw in their Buddhists and nuns.

Investigations of meditation's therapeutic value have been equally inconclusive. Meditation has been linked to a dizzying array of benefits, including the alleviation of stress, anxiety, high blood pressure, substance abuse, hostility, pain, depression, asthma, premenstrual syndrome, infertility, insomnia, and the side effects of chemotherapy.* But many of these studies have been poorly designed, Andreson remarked, carried out with inadequate controls or no controls at all.

---

*The world's leading scientific proponent of the healing benefits of meditation, prayer, and religious faith is Herbert Benson, a physician at the Harvard Medical School. He was the author or coauthor of sixty-three of the books and papers on meditation cited by Andreson. For a devastating — and amusing — critique of Benson's work, see "Mind and Body," by Irwin Tessman and Jack Tessman (*Science*, April 18, 1997, pp. 369–370). Irwin Tessman, a biologist, and his brother Jack, a physicist, reviewed *Timeless Healing*, by Benson and the journalist Marg Stark (Fireside, Simon and Schuster, New York, 1996). The Tessmans contended that

Andreson noted that meditation has been linked to adverse effects, too, including suggestibility, neuroticism, depression, suicidal impulses, insomnia, nightmares, anxiety, psychosis, and dysphoria. In an implicit reference to crazy-wisdom gurus such as Rajneesh, Trungpa, and Da Free John, Andreson added that meditators may become vulnerable to "manipulation and control by others," including "unscrupulous or delusional teachers."

To someone who has read Andreson's paper in the *Journal of Consciousness Studies,* Newberg and D'Aquili's article on the neurophysiology of meditative states in the same issue must seem comically simplistic. They declared that "feedback through the limbic structures to both the left and right PFC [prefrontal cortex] . . . results in instantaneous maximal stimulation of the left PFC with immediate total blocking of input into the left PSPL [posterior superior parietal lobe] which may be associated with obliteration of the Self-Other dichotomy." As one critic observed recently, Newberg and D'Aquili's model conjures up an image of the brain as a Rube Goldberg contraption spitting out neatly packaged "states of absolute unitary being."

Enlightenment, orgasms, and spandrels

Although aware of the limitations of his research, Andrew Newberg was understandably optimistic about its prospects. He said that many other researchers at the University of Pennsylvania — psychiatrists, psychologists, social scientists, criminologists, and others — have expressed interest in his work. The university is creating a center on spirituality, which will focus on the therapeutic potential of religious faith and experiences. He has re-

---

few of Benson's claims about the effects of meditation and other spiritual disciplines withstand scrutiny. Benson reported, for example, that he had observed Tibetan monks wrapped in damp sheets tolerate a temperature of 4.4 degrees centigrade without shivering; he claimed that the monks' "ancient meditation techniques" allowed them to endure this condition, which would make normal people "shiver violently." "We wondered," the Tessmans wrote. "So one of us (77 kilograms, 1.79 meters, 67 years) repeated the stunt . . . [T]his aged beginner experienced no violent shivering or even serious discomfort. None of this would surprise wacky Green Bay Packer fans cavorting bare-chested in wintry Lambeau Field." The Tessmans have also critiqued studies that purportedly prove you can help a person recover from illness by praying for her, even if she doesn't know she's being prayed for and thus is not subject to the placebo effect. See "Efficacy of Prayer," by Irwin and Jack Tessman (*Skeptical Inquirer,* March/April 2000, pp. 31–33).

ceived largely positive coverage from the media, including a cover story in *Newsweek*.

Newberg has also received funding from the Templeton Foundation, created by the financier and devout Christian John Templeton to promote reconciliation between science and religion. The foundation funnels millions of dollars annually into programs "forging stronger relationships and insights linking the sciences and all religions."* It no doubt hopes that neurotheology might provide an empirical underpinning for religious belief. Newberg and D'Aquili suggested as much in *Why God Won't Go Away.* The "existence of an absolute higher reality or power," they declared, "is at least as rationally possible as is the existence of a purely material world." But their neurotheology seems unlikely to satisfy anyone's spiritual yearnings.

The state of absolute unitary being, Newberg and D'Aquili said, reveals that beyond this vale of tears there lies a transcendent, absolute, intensely real, unified . . . something. Although they suggested that mind is somehow as intrinsic to reality as matter, Newberg and D'Aquili made no claims about whether consciousness is eternal or whether life persists after death. They did not assert that we are in good hands; the God they envision is not a personal or a knowable God. Their neurotheology is a kind of negative theology. It recalls what the Scottish philosopher David Hume once said of mysticism: "Or how do you Mystics, who maintain the absolute incomprehensibility of the Deity, differ from skeptics or atheists who assert that the first cause of All is unknown and unintelligible?"

Indeed, some of Newberg and D'Aquili's speculations seemed more compatible with atheism than with religious belief. In *Why God Won't Go Away,* they proposed that our mystical capacity may have evolved out of our orgasmic capacity. Both orgasms and mystical experiences, Newberg and D'Aquili pointed out, produce sensations of bliss, self-transcendence,

---

*The Templeton Foundation publishes a magazine, maintains a Web site, and provides grants for research projects, lectures, conferences, books, articles, academic courses, and other activities that further "progress in religion." One outspoken critic of the foundation is the physicist Steven Weinberg. In "A Designer Universe" (*New York Review of Books,* October 21, 1999, pp. 46–48) he wrote: "I am all in favor of a dialogue between science and religion, but not a constructive dialogue. One of the great achievements of science has been, if not to make it impossible for intelligent people to be religious, then at least to make it possible for them not to be religious. We should not retreat from this accomplishment." Ironically, in *Why Religion Matters* (HarperSanFrancisco, 2001), Huston Smith criticized the Templeton Foundation for being excessively deferential to science.

and unity; that may be why mystics such as Saint Teresa so often employed romantic and even sexual language to describe their raptures. Mystical experiences can be induced by rhythmic activities, as orgasms are; orgasms trigger the simultaneous activation of our arousal and quiescent systems, as unitive experiences do.

Newberg and D'Aquili noted that the overlap between mystical rapture and orgasms is not complete. A deep-rooted brain structure called the hypothalamus, which regulates both arousal and quiescent systems, seems to play a larger role in orgasms; the frontal lobes, the seat of higher cognitive functions, are apparently more active during mystical experiences. Nevertheless, they concluded, an "evolutionary perspective suggests that the neurobiology of mystical experience arose, at least in part, from the mechanism of the sexual response."*

This hypothesis reminds me of a 1979 paper by the Harvard biologists Richard Lewontin and Stephen Jay Gould called "The Spandrels of San Marco and the Panglossian Paradigm." Dr. Pangloss is a character in Voltaire's *Candide.* He is a self-described professor of metaphysicotheologicocosmonigology, an eighteenth-century precursor of neurotheology. This is the best of all possible worlds, Pangloss declares, and everything in it serves a human purpose: Our noses were perfectly designed to bear up our spectacles, pigs to keep us fed year-round, stones to make castles from, and so on.

In their paper, Lewontin and Gould accused some evolutionary theorists of fallacious Panglossian reasoning. The neo-Panglosses argue that virtually all human traits were originally instilled in us by natural selection because they served some purpose, namely, helping our ancestors propagate their genes; in evolutionary parlance, these functional traits are adap-

---

*More than a century ago, William James felt compelled, in *The Varieties of Religious Experience,* to deplore "the fashion, quite common nowadays among certain writers, of criticizing the religious emotions by showing a connection between them and the sexual life" (p. 27). Religious experience arguably depends "just as much upon the spleen, the pancreas, and the kidneys as on the sexual apparatus" (p. 29). But some spiritual traditions have encouraged this linkage of sexuality and mysticism. Kundalini yoga and tantric yoga both seek to harness sexual energy in the service of mystical awakening. Tibetan Buddhism holds that we can glimpse the clear light of the void not only during meditation and sex but also while sleeping and sneezing (*Sleeping, Dreaming, and Dying,* edited by Francisco Varela, Wisdom Publications, Somerville, Massachusetts, 1997, p. 49). A spiritual seminar comes to mind: "Snooze and sneeze your way to enlightenment!"

tations. According to Lewontin and Gould, many modern human faculties probably originated not as adaptations per se but as incidental *byproducts* of adaptations. They called these nonadaptive features "spandrels" — an architectural term for the triangular space between an arch and its surrounding structure. The spandrel, at least originally, serves no function itself; it is merely a byproduct of the truly functional feature, the arch.

Applying this terminology to Newberg and D'Aquili's conjecture about the biological origin of mysticism, the orgasm is clearly an adaptation, because it provides a powerful motivation for sexual congress and hence gene propagation. Through some evolutionary quirk, perhaps, our orgasmic capacity gave rise to our mystical capacity. If orgasms are arches, then mystical experiences — which perennial philosophers from Richard Bucke to Ken Wilber have implied are the telos of evolution — are nothing but spandrels.

Just because our mystical capacity is a spandrel doesn't mean it cannot yield deep insights into reality. The human capacity for science is arguably a spandrel, an incidental byproduct of our innate curiosity and capacity for pattern recognition. Science has nonetheless proved to be an enormously powerful method for discovering truth. The point is that the evolutionary origin of our mystical capacity has no bearing on the truth and value, or lack thereof, of mystical visions. As William James pointed out, such experiences must be judged by their fruits, not their roots. Moreover, if a spandrel is defined as something purposeless and unnecessary, then *everything* is arguably a spandrel, including consciousness, life, the entire universe.

Temporal-lobe syndrome

As Newberg would be the first to admit, his investigations represent only a tentative first step toward understanding mysticism; his hypothesis about mysticism's neural basis might have to be revised or abandoned as research proceeds. In fact, some of Newberg's fellow neurotheologians have focused on other neural regions, including the temporal lobes, two large structures that wrap around the brain beneath each ear, and the limbic system, a network of structures nestled deep in the temporal lobes. Newberg has not ignored the limbic system, which regulates our emotional responses; he suggests that limbic excitation may give rise to the powerful emotions felt by mystics. But the *sine qua non* of his model is the posterior superior parietal lobe, which supposedly mediates the sense of unity felt by mystics.*

One neuroscientist who has given more prominence to the temporal lobes and limbic system is V. S. Ramachandran of the University of California at San Diego, who specializes in brain damage that leads to peculiar disruptions in our perception of reality and of ourselves. In his 1998 book *Phantoms in the Brain,* cowritten with the science journalist Sandra Blakeslee, Ramachandran discussed such bizarre disorders as Capgras's syndrome, the delusion that people around you have been replaced by impostors, as in the film *Invasion of the Body Snatchers;* anosognosia, the inability of someone paralyzed on one side of the body to perceive the paralysis; and phantom-limb syndrome, in which someone continues to receive sensory signals from, or "feel," an amputated limb.

By studying disorders such as these, Ramachandran said, neuroscientists can learn much about how the brain constructs reality. He has called his approach "experimental epistemology." In a chapter titled "God and the Limbic System," Ramachandran turned his attention to religious and mystical visions. He stressed that his intention is not to denigrate mysticism or religion — or to suggest that they are pathologies stemming from brain disorders — but only to understand them. Ultimately, God's existence or nonexistence is beyond the purview of science.

The temporal lobes, Ramachandran noted, and particularly the left temporal lobe, have long been implicated in religious visions: "Every medical student is taught that patients with epileptic seizures originating in this part of the brain can have intense, spiritual experiences during the seizures and sometimes become preoccupied with religious and moral issues even during the seizure-free or interictal periods."

Recently, Ramachandran had encountered a patient who was a "textbook case" of this so-called temporal-lobe syndrome. Paul was a thirty-two-year-old epileptic who obsessively talked and wrote about religious subjects. He told Ramachandran that during his epileptic seizures he felt a

*Recent studies of monkeys and brain-damaged humans have undercut a major premise of Newberg and D'Aquili's model, that the posterior superior parietal lobe helps us orient our bodies in the external environment (and hence that decreased activity within the lobe generates sensations of mystical oneness). See "Spatial Awareness Is a Function of the Temporal Not the Posterior Parietal Lobe," by Hans Otto Karnath, Susanne Farber, and Marc Himmelbach (*Nature,* June 21, 2001, pp. 950–953).

rapturous "Oneness with the Creator" that carried over into the rest of his life. When Ramachandran asked Paul if he believed in God, Paul replied, with a puzzled expression, "But what else *is* there?"

Ramachandran offered a tentative guess that the left temporal lobe might be the "God module," the neural circuit that underpins our deepest spiritual sentiments. Obviously not all those who report mystical and religious visions suffer from epilepsy, but Ramachandran suspected that elevated neural activity in the left temporal lobe might contribute to the experiences of healthy mystics, too. As further evidence for this hypothesis, Ramachandran cited reports about a device that supposedly induced mystical experiences in normal subjects by stimulating their temporal lobes with electromagnetic pulses. Some journalists were calling the device "the God machine."

5

The God Machine

I FIRST HEARD about the God machine in 1997 during a visit to Amsterdam. When I mentioned my interest in mysticism to a Dutch journalist, she said she had recently seen a television show about a device that can induce religious experiences by electromagnetically stimulating the brain. The journalist couldn't remember the name of the device's inventor, but she promised to send me the information. Shortly after I returned to the United States, she informed me by e-mail that the inventor was Michael Persinger, a psychologist at Laurentian University in Sudbury, Canada.

When I did a Web search on Persinger, I found that he is an ambitious researcher, to put it mildly. The neurotheologian Andrew Newberg is content to posit a neural mechanism for one mystical condition: the state of absolute unitary being, in which our sense of self and the outside world dissolves into formless unity. Persinger has proposed a grand unified theory of virtually all altered states, and he has developed a machine for inducing them.

Underlying Persinger's work is his conviction that anomalous electromagnetic fluctuations — produced by solar flares, seismic activity, radio and microwave transmissions, electrical devices, and other external sources, or originating in the brain itself — can trigger disturbances resembling epileptic seizures. These "microseizures," he proposes, generate a wide range of altered states, including religious and mystical visions, out-of-body experiences, and even alien-abduction episodes.

Persinger conjectures that our sense of self is ordinarily mediated by the brain's left hemisphere — more specifically by the left temporal lobe. When the brain is disrupted — by a head injury, epileptic seizure, stroke, drugs, psychological trauma, or external electromagnetic pulses — our left-brain self may detect activity in the right hemisphere as another self, or what Persinger calls a "sensed presence." Depending on our circumstances and background, we may perceive the sensed presences as extraterrestrials, ghosts, angels, fairies, muses (that is, sources of artistic inspiration), demons, or God Almighty.

In the late 1980s, Persinger began testing this theory on healthy volunteers with a device that stimulates the brain. The latest version of his invention, which Persinger calls the Octopus, consists of eight electrified magnets, or solenoids, that encircle the head and deliver computer-controlled electromagnetic pulses to specific regions of the brain. Persinger claims that as many as 40 percent of his subjects "sense a presence" while they are being stimulated, compared to 15 percent of a control group.

Persinger's work has been publicized by ABC News, *Newsweek* (in the same article that highlighted the work of Andrew Newberg), *Wired,* and many other mass-media outlets. Some of the coverage has been lurid. One article called Persinger's brain-stimulating device "a radio for speaking to God," echoing the description of the Ark of the Covenant in the film *Raiders of the Lost Ark.* An article published by the Canadian magazine *Maclean's* apparently coined the phrase "the God machine."

As bizarre as it sounds, Persinger's research builds on a large body of empirical observations by others. The linkage of mystical states and epileptic seizures dates back at least to the ancient Greeks, who called epilepsy a "sacred disease." Medical historians have suggested that some prominent religious visionaries — including the prophet Ezekiel, Saint Paul, Muhammad, Joan of Arc, the Swedish mystic Emanuel Swedenborg, and Joseph Smith, the founder of Mormonism — suffered from seizures.* Various more modern figures — notably Fyodor Dostoyevsky — have re-

*The linkage of temporal-lobe seizures and mystical visions inspired Mark Salzman's elegant novel *Lying Awake* (Knopf, New York, 2000). Salzman's heroine is Sister John of the Cross, a modern-day Carmelite nun living in a convent near Los Angeles. After years of searching fruitlessly for God in her prayers, she begins having fainting spells, during which she feels herself dissolving into a brilliant, blissful radiance that she believes — she *knows* — is God. When her seizures become more frequent, the prioress persuades Sister John to see a neurologist, who finds a tumor growing in her temporal lobe. Salzman's novel delicately probes — but makes no attempt to resolve — the vexing questions that arise when spirituality collides with science: If a mystical vision can be traced to a brain tumor or some other ailment, does that mean it is false? Or are *all* mystical experiences delusions, generated not by God but by our own brains?

ported that their seizures were preceded by a flash of mystical rapture. A century ago the theory that epileptic seizures cause mystical states was so popular that William James felt compelled to denounce it as simplistic.

Persinger's other ideas have precedents as well. In his 1976 book *The Origin of Consciousness in the Breakdown of the Bicameral Mind,* the psychologist Julian Jaynes proposed that tens of thousands of years ago the corpus callosum, the bundle of nerves connecting the brain's hemispheres, was much less developed than it is today. As a result, the left hemisphere, the primary seat of language and identity, commonly attributed signals emanating from the right hemisphere to an external source, such as a ghost or a god. Only when the two hemispheres became better integrated did the relatively unified consciousness typical of modern humans emerge, according to Jaynes.

In support of his theory, Jaynes cited experiments by the Canadian neurosurgeon Wilder Penfield on epileptic patients in the 1950s. While preparing the patients for brain surgery, Penfield stimulated different regions of their brains with electrodes and asked the patients to describe any sensations. (Because the brain does not contain any pain receptors, patients undergoing brain surgery need not be put to sleep with a general anesthetic.) Some patients whose right temporal lobes were stimulated reported hearing voices and seeing apparitions. After hearing about Penfield's experiments, Aldous Huxley wrote in his essay "Heaven and Hell": "Is there, one wonders, some area in the brain from which the probing electrode could elicit Blake's Cherubim?"

Michael Persinger's work is inspiring scientists to ask similar questions. V. S. Ramachandran — who shares Persinger's suspicion that the temporal lobes underpin spiritual experiences — has pondered whether the God machine could turn atheists into religious believers. Ramachandran mused that his friend Francis Crick would make an excellent test case. Best known as the codiscoverer of DNA's double helix, Crick is a notorious scientific skeptic. He once asserted that the supposedly immortal, immaterial "soul" postulated by Christianity and other religions is "nothing but a pack of neurons," and he has speculated that belief in God is induced by mutant neurochemicals called "theotoxins."

In the Home of the Big Nickel

Eager to see how the God machine would affect my skeptical brain, I flew to Toronto in the spring of 1999 and caught a puddle-jumper north to Sudbury. A mining town dominated by the International Nickel Company, Sudbury has all the charm of a Siberian gulag. The landscape is flat except for eruptions of black, slaglike rock. Vast industrial structures smolder on the horizon. The town's tourist attractions include "the Big Nickel," a coin twenty-seven feet across and two feet thick, and "the world's tallest smokestack." The route from the Sudbury airport to Laurentian University wound through bland apartment buildings draped with signs pleading for renters. Painted on the side of one building was this message: "A train of thought is only as good as its engineer." The Laurentian campus resembled an industrial park. As instructed, I waited for Persinger in the lobby of the utterly unaesthetic Arts Building. He arrived at precisely 3:00 P.M., as we had agreed, accompanied by a bearded young graduate student.

Persinger was trim, ramrod straight, with close-cropped silver hair. He wore a three-piece pinstriped charcoal suit, a white shirt with French cuffs and cufflinks, a black-and-white-checked tie, and gold-rimmed aviator glasses. A colleague of his told me later that he had never seen Persinger dressed other than in a three-piece suit, even when he was mowing his lawn. Draped across Persinger's solar plexus was a gold chain attached to a watch tucked into his vest pocket. Whenever Persinger pulled the watch out and consulted it, for some reason I thought of the White Rabbit in *Alice in Wonderland.*

Persinger's voice and manner were as parodically formal as his attire. No first names. He and his graduate student called each other Mr. Cook and Dr. Persinger; I was Mr. Horgan. Persinger went over our schedule with the precision of a military commander. It is now 3:07. We will talk until 4:00. Then I will have my session with the Octopus, which will last approximately one hour. After that, we can talk again until 5:30, when Persinger has another engagement. If I like, we can also speak again from 10:00 to 11:30 P.M. When I marveled that he worked so late, Persinger informed me that he would be working well after 11:30; he could talk to me only until that time, and then he had to attend to other tasks.

Persinger is not a native of Canada. He moved there in 1969 after getting a master's degree in physiological psychology from the University of

Tennessee. He obtained a doctorate from the University of Manitoba in 1971 and came to Laurentian shortly thereafter. He reveled in his role as scientific maverick. "People can criticize me and make fun of me and laugh at me. As long as they don't try to kill me, I'm going to keep doing what the data tell me are important." Persinger seemed almost proud that he had received only one grant for his research, back in 1983. He paid for his research with income from his private practice, which involves treating people with head injuries.

His clinical practice helped to inspire his interest in the neurology of religious belief. Many brain-injured patients intuit what Persinger calls "sensed presences," although few admit it without prodding. "They think they're going crazy," he said. If the injury is in the left hemisphere, the presence may emerge as a voice; the experience is also more likely to be pleasant or even ecstatic. Experiences stemming from injuries to the right hemisphere are more likely to be frightening; patients may think they are being haunted by an evil ghost or demon. The patient's personality and background also determine how he or she interprets the experience.

Persinger insisted that his hypotheses are not intended to undermine religious belief by reducing it to mere neurophysiology or neuropathology. "I'm not trying to dismiss God," he said. "I'm trying to understand the area of the brain that's involved in the electrical pattern" that generates religious belief. Persinger contended that God's existence can be neither proved nor disproved by science. "It's something that is invisible, nonphysical, eternal, and everywhere. If you're going to define it that way, then there's no way you could ever test it."

In spite of his professed neutrality, Persinger is clearly less sympathetic toward either religion or mysticism than neurotheologians such as Andrew Newberg and Eugene D'Aquili. In his books *TM and Cult Mania* and *Neuropsychological Bases of God Beliefs*, published in the 1980s, he cast religious belief and spiritual practices in a psychopathological light. In a 1993 paper he argued that those who respond strongly to meditation could be suffering from a kind of low-grade epilepsy. Persinger admitted to me that he is not a great fan of religion. Religious institutions have caused a great deal of harm, he said, particularly when they insist that believers are superior and nonbelievers are inferior or evil. "That's where religion becomes a damaging force."

If perfected, the technology underlying the God machine could be used for ill or good, Persinger said. On the one hand, it might induce blissful mystical visions in depressed or dying people more reliably and safely than psychedelic drugs. But such a technology could also be harnessed for insidious purposes. "Just think of the practical impact" of a device that can implant beliefs and commands that seem to come straight from God. "That's the highest motivation there is. People will die for this."

The key to perfecting this mind-control technology, Persinger said, is finding the code or language whereby brain cells store and transmit information. The shape and timing of an electromagnetic signal, rather than its amplitude, determines its ability to influence a brain. Similarly, I was responding to the content of his words, Persinger explained, rather than to their loudness. "Once you understand me above my whisper," he said, lowering his voice until I could scarcely hear him, "it really doesn't make any difference except for convenience on my part whether I do it at thirty or ninety decibels."

Strapping on the Octopus

During my session with the Octopus, I would be exposed to two different sets of signals that Persinger suspected can generate sensed presences. The first set is based on electromagnetic oscillations emitted by the amygdala during certain epileptic seizures. The second set is based on an electrical pattern that triggers a seizurelike firing of brain cells isolated in a petri dish. The amplitude of the signals is deliberately kept very low, Persinger said, comparable to those produced by a hair drier or other household appliance.

"When we designed this, the last thing we wanted to do was have vivid hallucinations, because that means you are introducing unbelievable energies into the brain that are potentially destructive. So what we try to do is to change experiences with the same level of modality as natural occurrences, so that they are a part of you. You believe they are real." In other words, subjects influenced in this way should not realize they are being influenced. "That's the *real* mind control," Persinger said.

My anticipation grew when I filled out a questionnaire designed to determine whether I was at risk for a mental meltdown in the God machine. Have I ever had any type of epileptic seizure? Psychosis? Do I ever see strange flashing lights? Smell odd odors? Hear voices? I signed a release form absolving Persinger and his helpers of all blame for anything that might happen. Then it was time to enter the chamber.

It was a soundproof cube, perhaps eight feet on a side. Green-painted particleboard lined the walls; a yellow shag carpet covered the floor. I sat in, or rather sank into, a ratty overstuffed armchair. Across from me stood two ancient speakers. The Octopus consisted of a Velcro headband covered with wires and film canisters, each of which contained a solenoid. Persinger's graduate student, Mr. Cook, fitted this contraption around my head, as well as goggles with tissue paper stuffed in the sockets. After some final fiddling, Cook left the chamber.

The room was echoless, muffled, cozy. I felt unusually peaceful and relaxed. As someone who works at home with two young children and two garrulous parrots, I was not used to this silence. The soundproofing wasn't completely effective; faint rustling sounds came from behind me as Cook took his place at a computer console just outside a window. I waited, hypervigilant, for something to happen. I heard a faint buzz, seemingly from the apparatus on my head, and sensed a gentle pulsing from my heartbeat or some other source. I saw vaguely geometric shapes, luminescent greens gliding over a black background.

Energy seemed to ebb and flow, surflike, through me. Induced or natural? I couldn't decide. There was nothing so abrupt or odd that I could with certainty blame it on the Octopus. There was only this soft, patterned darkness and my incessant, dithering internal monologue. Am I feeling something? Is this unusual? I came all the way up here to this godforsaken place and nothing is happening? How will I turn this into a scene for my book?

Abruptly, a peculiar memory intruded on my thoughts. It dated back to the late 1970s, when I was living in Denver and a friend persuaded me to join her in taking an adult-education class called "You and Me and ESP." The instructor, who favored baby-blue leisure suits and a glossy pompadour, assured us that we all have innate psychic abilities, we just have to *buuh-leeeve* in *ow-uh pow-uh!* He also said that each of us has a spiritual "other," a kind of guardian angel with whom we can communicate if we try hard enough.

I didn't take this notion seriously enough to act on it until one evening when a couple of friends and I visited a sensory-isolation parlor. The establishment contained a dozen or so coffinlike tanks filled with salt water kept at skin temperature. (Sensory-isolation tanks were invented in the 1950s by the neuroscientist John Lilly.) Floating in my tank, I idly decided to try to contact my spiritual other. After a minute or two someone, something, rapidly approached me from deep space. Panicked, I tried to repulse this being, to wish it away. When it kept approaching, I threw open the door of the tank and lurched out, hyperventilating.

Musing over this memory in Persinger's cubic chamber, I realized how unlikely such an experience would be now. I had grown more skeptical over the past twenty years, and as Persinger warned, those who are incredulous of the God machine usually have uninteresting experiences.

"John?" It was the voice not of God but of Persinger's graduate student, Mr. Cook. The experiment was over. No demons, ghosts, guardian angels, or deities. Only the effluvia of my skull-sealed, skeptical brain.

Mind control and the neural code

Actually, Persinger's work merits skepticism, for several reasons. Media coverage of Persinger often implies that his subjects have profound religious visions — of "Elijah, Jesus, the Virgin Mary, Mohammed, the Sky Spirit," as *Wired* put it. But his papers describe much less dramatic results. Mr. Cook, who'd supervised hundreds of Octopus sessions in the late 1990s, told me that most subjects who sense a presence usually just have a vague feeling of being watched, with little emotional affect — and of course they *are* being watched, by the researchers. Cook could not recall anyone attributing divine or demonic qualities to the sensed presence. The experience is commonly described as "relaxing," "interesting," or "really neat," he said.

Some subjects do report odd sensations. A Canadian journalist sensed a "cold" presence in the chamber with him and saw a pair of eyes staring at him. After my session, I met the journalist and learned that he had first seen this "ghost" — his word — in his own home long before entering Persinger's chamber. The writer Ian Cotton, who specializes in religious topics, daydreamed that he was "in a temple, in a line of solemn, Tibetan monks, grave-eyed, brown cowls around their heads, the bells tolling loudly now, echoing in my head." He acknowledged that his experience was probably triggered by the "vaguely New Age, Eastern temple bell sounds" emitted by the speakers in the chamber. No one stimulated by the God machine has publicly reported anything resembling a classic mystical experience, with sensations of bliss, unity, and ineffability.

In *Phantoms in the Brain,* V. S. Ramachandran stated — or misstated, rather — that Persinger "experienced God" while stimulating his own temporal lobes with the God machine. Persinger has tried out his own invention, but he assured me that he has never had any religious or mystical sensations, either while wearing the Octopus or in any other context. He suspected that his rational, skeptical outlook inhibited him. (So much for Ramachandran's speculation that the God machine might turn atheists such as Francis Crick into believers.) In other words, Persinger's experiments do not really provide much evidence for the theory that temporal-lobe excitation can cause religious and mystical visions in healthy subjects.

Other evidence for this theory is weak as well. Ramachandran has acknowledged that the evidence linking temporal-lobe activity with mysticism and "hyperreligiosity" in epileptics is largely anecdotal and disputed by many neurologists. Other scientists agree that few temporal-lobe epileptics — who account for more than one third of the two million epileptics in the United States — report mystical or religious visions during their seizures. Conversely, most people who have had mystical experiences show no signs of epilepsy. That is not to say that the linkage of temporal-lobe excitation and mysticism is totally spurious. Unusual activity in the temporal lobes might account for some mystical visions — including those of Saint Teresa and other famous religious figures — but it cannot account for all of them.

Persinger's most portentous claim by far is that his Octopus device may be the prototype of a truly powerful and precise mind-control machine. Of course, technologies already exist for modifying human cognition with external electrical stimulation of the brain. One obvious example is electroshock therapy, which in spite of its poor reputation has recently made a comeback as a treatment for severe depression and other disorders. It jolts the brain with doses of electricity large enough to initiate an artificial seizure or convulsion (hence its common synonym, electroconvulsive therapy). Researchers are now testing whether much milder pulses of electricity, comparable to those that the God machine delivers, can treat depression and other illnesses.

These technologies are primitive compared to what Persinger envisions. In a 1995 paper titled "On the Possibility of Directly Accessing Every

Human Brain by Electromagnetic Induction of Fundamental Algorithms," he declared that scientists are on the verge of deciphering the "algorithms" by means of which our brains process information. Once scientists achieve this feat, they will be able both to monitor and to control human thoughts remotely with electromagnetic signals — or as Persinger put it, "to influence directly the major portion of the approximately six billion brains of the human species without mediation through classical sensory modalities."

This prophecy is far-fetched, to say the least, given the current state of neuroscience. First of all, developing technology that can control minds remotely would require solving what is arguably the most difficult problem science has ever faced: deciphering the neural code, the language whereby brain cells gather, store, and communicate information. Some scientists believe that the code may be too complex to decipher. There may in fact be innumerable neural codes governing the countless tasks our brains carry out.

Even if researchers someday discover a single neural code — perhaps comparable in its simplicity and universality to the DNA-based genetic code — that finding is extremely unlikely to yield the remote mind-control technology that Persinger envisions. The electromagnetic "algorithms" with which he stimulates brains are based on EEG readings. Through electrodes placed on the scalp, EEG machines detect electrical fluctuations resulting from the firing of billions of neurons.

EEG signals can indicate the general state of a brain, revealing whether someone is alert or drowsy or dreaming or in the midst of an epileptic seizure. But EEG signals are more distant from any conceivable neural code than the roars of a crowd in a stadium are from human language. The only way to manipulate the brain *precisely* — to speak to it in its own language, as it were — would be through vast numbers of implanted electrodes that communicate directly with individual neurons.

Such a technology seemed more feasible half a century ago, after Wilder Penfield and other researchers showed that implanted electrodes can trigger intense emotional and behavioral reactions. In the early 1950s, the neuroscientist Robert Heath of Tulane University induced intense pleasure in psychiatric patients with electrodes implanted in the septum, a minute region just above the hypothalamus. Heath's group also artificially induced multiple orgasms in a female patient by injecting the neurotransmitter acetylcholine directly into her septal region.

In similar experiments, José Delgado of Yale University showed that he could induce pleasure, rage, and other emotions in psychiatric patients with implanted electrodes and could control patients' bodies. In an exhibition in a bullring in Spain in 1963, Delgado stopped a bull charging toward him by radioing a signal to an electrode implanted in its brain. He carried out similar tricks with psychiatric patients. In 1969 Delgado prophesied, in *Physical Control of the Mind,* that before long implanted devices might eliminate human violence and other maladaptive behaviors.

These fantasies have receded, in part for obvious ethical and legal reasons. It also became clear, even to enthusiasts like Heath and Delgado, that implanted electrodes could at best exert only crude control over emotions and behavior, and at a great risk to mental and physical health. Implanted electrical-stimulation devices are now occasionally used for treating epilepsy, severe tremors, and other disorders. Given the unpredictability of their effects, however, and the risks of brain surgery and implantation, these technologies are used only as measures of last resort.

Psi, God, and pattern recognition

I have another, perhaps excessively idiosyncratic reason to doubt Persinger's credibility. My recollection during my Octopus session of my ESP class in Denver turned out to be weirdly prescient. As I pulled the gear off my head at the end of the session, Mr. Cook entered the chamber and handed me five sheets of paper with different images printed on them: a herd of zebras, two men climbing what appeared to be a wall of mud, a crowd of men fighting, fish drifting through green-blue water, an elderly woman. Cook asked me to rank the pictures according to how closely they resembled my visual sensations. Puzzled by this exercise, I ranked the underwater scene first and the fighting men last.

With a small smile, Cook told me that he had been testing one of Dr. Persinger's pet theories. Persinger suspected that extrasensory perception might be enhanced by sensory deprivation combined with electromagnetic stimulation. While I was being stimulated by the Octopus, Cook had been staring at one of these images, trying to imprint it on my mind. The image he had been staring at was the underwater scene. At that point Persinger entered the chamber. Cook waved the underwater picture at him and said, "We got a hit." Persinger nodded coolly, as if he had expected nothing less.

I was taken aback. I told Persinger and Cook that their remote-viewing experiment didn't impress me. I had chosen the underwater scene because it was the most abstract and dreamy of all the images. It would have been a better test if Cook had tried to transmit one of the other pictures, which were more vivid and detailed. I told Persinger that I don't believe in ESP, and I was surprised to learn that he does.

Persinger retorted that he neither believes nor disbelieves; he is an *investigator.* He is trying to gather data, to see if something is there. "The critical thing is, can you get a reliable phenomenon?" So far, subjects were choosing the correct image about 75 percent of the time; one would expect only a 20 percent rate according to chance. To ignore these data, Persinger declared, "would be just as irresponsible scientifically as someone saying, 'Oh yeah, that's all true.' "

I had assumed Persinger was, if anything, a debunker of paranormal phenomena; that is how he has been portrayed by psi doubters such as Michael Shermer, publisher of *Skeptic* magazine. In spite of his demurrals about scientific neutrality, Persinger has clearly suggested that alien-abduction reports and religious and mystical visions stem from brain-induced delusions. But by claiming that subjects were choosing the correct image 75 percent of the time in his remote-viewing experiments, he was also suggesting that extrasensory perception is a genuine phenomenon.

This episode forced me to confront the linkage of mysticism and psychic phenomena, or psi. Back in the late 1970s, when I attended the class "You and Me and ESP," if you had asked me whether I believed in psi, I probably would have shrugged and pleaded agnosticism. I had never had a clear-cut psychic experience, but I had friends who swore they had — including the woman who persuaded me to join her in taking the class. Over the next decade, my years as a science writer turned me into a hard-core skeptic. I came to view psi as a byproduct of our brain's ability to find patterns whether they exist or not. I saw belief in and even agnosticism toward psi as symptoms of intellectual flakiness.

When I began researching this book, I thought I could investigate mysticism without considering psi, but I soon realized just how closely mysticism has always been linked to miracles, magic, and other supernatural phenomena. Jesus, after his mystical sojourn in the desert, supposedly raised Lazarus from the dead and turned water into wine. Many modern gurus, too, reportedly possess psychic abilities: telekinesis, extrasensory perception, the power to heal. Sai Baba, a popular Indian guru, conjures up watches, gems, candy, and other items seemingly out of thin air. Advocates of Transcendental Meditation offer courses in levitation and claim that they have reduced crime rates with their thought waves.

William James noted that the term *mystic* often refers to someone who believes in supernatural phenomena such as "spirit-return" and "thought-transference" — that is, in ghosts and telepathy. Although James excluded these occult connotations from his definition of mysticism, he did not rule out the existence of supernatural phenomena. His contemporary Richard Bucke went even further, claiming that as humanity's mystical awareness evolves, so will its ability to perform "supranormal" feats such as telepathy and telekinesis.

A surprising number of more recent scholars — notably Aldous Huxley and Huston Smith — have also suggested that mystical knowledge may awaken latent paranormal capacities. Ken Wilber, as eager as he is to project a scientifically conservative image, once stated, "I'm sure [psychic phenomena] exist." He hastened to add that psychic phenomena "don't interest me very much, and in any event they have little to do with mysticism per se."

Actually, belief in extrasensory perception, telekinesis, and other psychic effects is a logical consequence of the mystical doctrine — promulgated by Wilber himself — that mind rather than matter constitutes the primary level of reality. As Bucke put it in *Cosmic Consciousness,* the cosmos is not "dead matter governed by unconscious, rigid, and unintending law" but is "entirely spiritual." Some mystics are content to express this mind-over-matter concept as an abstract metaphysical principle. Others assert that when your cosmic consciousness awakens, you can attain virtually infinite power over the material realm, in the form of clairvoyance, astral projection, and other occult capacities.

As my research of mysticism continued, I encountered so many intelligent, reasonable people present and past who believe in psi or do not rule it out that I was forced to review my own stance. Paranormal phenomena, it seems to me, can be explained in three basic ways. One explanation is that these phenomena are miracles, divine overrides of nature's laws. According to this view, certain people receive special treatment from God because of their special relationship to Him. God parts seas for them, changes water into wine for them, raises them from the dead, presumably because

He likes them more than the rest of us. Because this explanation is tainted by the scandal of particularity, I find it the hardest to accept.

The second, slightly more palatable view — often espoused by New Age authors — holds that telekinesis and clairvoyance are natural phenomena that will someday be explainable by still undiscovered laws of nature, possibly related to quantum mechanics. We all have psychic powers; some of us are just better than others at exercising them. In the future, science may help more of us learn to exploit our latent psychic capacities. The problem with this second explanation is that psi has never been convincingly demonstrated in the laboratory, the claims of Persinger and many others notwithstanding.

That leads to the third explanation for psi, that it stems from our compulsion for extracting meaningful patterns from the chaos of the world. In *Why God Won't Go Away*, Andrew Newberg postulated the existence of an innate cognitive function called the "causal operator," which compels us to find causes and effects for events. The causal operator can help us discover genuine cause-and-effect relationships — embodied, for example, in the laws of physics — as well as spurious ones. According to Newberg, studies have shown that people insist on finding a pattern in a sequence of events even when they are told that the sequence is entirely random.

We seem to be especially prone to detecting an intelligence or mind guiding events both large and small. The anthropologist Stewart Guthrie of Fordham University has proposed that our religious intuitions stem from our instinctive tendency to project human qualities onto the world. Our anthropomorphism is an inborn, adaptive trait, Guthrie contended, which enhanced our ancestors' chances of survival. "[I]n the face of chronic uncertainty about the nature of the world, guessing that some thing or event is humanlike or has a human cause constitutes a good bet," he wrote. This strategy is "the product of natural selection, not of reason."

If a Neanderthal mistook a tree creaking outside his cave for a human assailant, Guthrie argued, he suffered no adverse consequences other than a moment's panic. If the Neanderthal made the opposite error — mistaking an assailant for a tree — the consequences might have been dire. Over time, as natural selection bolstered our anthropomorphic tendency, it extended beyond specific objects, events, and phenomena to all of nature. We interpreted even the most random-seeming natural phenomena as the acts of a wrathful or loving deity. According to Guthrie, religious and mystical visions of divinity underlying reality are just extreme manifestations of our compulsive anthropomorphizing.

A similar theory has emerged from a field called evolutionary psychology, which views the human mind as a collection of adaptations sculpted by natural selection in our primordial past. Some evolutionary psychologists have speculated that soon after infancy all healthy children manifest an innate, intuitive ability to infer the state of mind of other humans. This awkwardly named "theory of mind" capacity has obvious survival value, because it helps us predict the actions of others for our own benefit. Damage to this capacity may cause autism. Autistics often seem incapable of distinguishing between humans and inanimate objects.

Many, perhaps virtually all, healthy humans have the opposite problem: an overactive theory-of-mind capacity. We unthinkingly discern sentience and complex psychological states not only in other humans and in animals but also in rainstorms, droughts, and shooting stars. Hence religion. But if we decide that supernatural phenomena — telekinesis, clairvoyance, ghosts, gods, and, yes, demiurges — are all illusions generated by our overactive causal and anthropomorphic operators, must we conclude that all our mystical insights are illusions as well? In other words, is it possible to be both a skeptic and a mystic?

6

The Sheep Who Became a Goat

Susan Blackmore is an extravagantly unconventional scientist. I met the British psychologist one frigid winter morning at an inn in Cambridge, Massachusetts, where she had stopped on a lecture tour. Her hair was orange, red, and yellow, dark-rooted, cut short as a boy's, with sideburns plunging like daggers past each multiringed ear. Words spewed from her pell-mell, accompanied by vigorous hand movements and facial expressions. She was keen on onomatopoeic sound effects: *Ahhhhh* (to express her pleasure at finding other smart people when she entered Oxford), *DUN da la DUN da la DUN* (the galloping noise she heard as she sped through a tree-lined tunnel in her first out-of-body experience), *Yan yan yan* (the sound of trees whipping past her as she sped through that tunnel), *Zzzzzzt* (the sound of reality dissolving after a toke of dimethyltryptamine), *Ooooooo* (how she felt after her first kensho).

She was easily distracted. When we spoke in the dining room of her inn, twice we had to move to a quieter spot when employees or patrons of the inn started talking near us. One side effect of her spiritual practice, she explained, is that she has a hard time blocking out distractions. "I think it is one of the bad effects of practicing mindfulness. I'm so *aware* of everything all the time."

The psychologist Theodore Barber coined the phrase "fantasy-prone personality" to describe people who have a higher-than-average tendency to report mystical, religious, and paranormal experiences. The fantasy-prone have vivid dreams and daydreams, and they are highly suggestible, responding strongly to hypnosis and guided imagery. Barber estimated that 4 percent of the population, mostly female, have these traits. I suspect that Susan Blackmore has a fantasy-prone personality. What makes her so unusual is that her visionary capacity is subject to an even more powerful skepticism and sense of reason.

Inspired by a spectacular out-of-body experience in college, Blackmore became a leading investigator of altered states, particularly those involving apparently paranormal phenomena. Gradually her research led her to conclude that all ostensibly supernatural phenomena have natural explanations. She denies the existence not only of ESP and astral projection but also of God, life after death, and free will. And yet Blackmore, who in 2001 left the University of the West of England in Bristol to become a freelance writer and lecturer, is a Zen Buddhist with a profoundly spiritual view of existence. "I see it as perfectly logical," Blackmore has written of her dual outlook, "but I know others find it confusing."

Blackmore makes a point of examining altered states from the inside as well as the outside. She has undergone "hypnotic regression" aimed at getting her to relive past lives, taken the anesthetic ketamine to test its capacity for inducing out-of-body experiences, and trained herself to have lucid dreams. She once strapped on Michael Persinger's God machine. When she did, she felt flashes of anger and fear, and her leg jerked as if pulled by a string. She came away convinced that excitation of the temporal lobes might indeed trigger exotic altered states, including hallucinations of aliens.*

As an undergraduate at Oxford in the early 1970s, Blackmore smoked pot, dropped acid, studied psychology, and joined several extracurricular groups, including the Society for Psychical Research, which is devoted to paranormal phenomena. She joined in part because the society's only other member on campus, Kevin, had long, curly hair. "I love men with long, curly hair." She glanced at my head and winced. "Sorry. Bad luck." Blackmore and Kevin played with Ouija boards and tarot cards; did experiments in telepathy; studied theosophy, the occult philosophy popularized in the late nineteenth century by the Russian-born guru Madame Blavatsky; and invited mediums and psychics to give talks at Oxford. Blackmore was struck by the contrast between the psychics' worldview and that of her

---

*Blackmore was surprised when I told her about the telepathy experiment Persinger had conducted with me. Perhaps knowing her reputation as a psi apostate, Persinger did not tell her about his sympathy toward psi when she visited him. Blackmore told the story of her encounter with Persinger's machine in "Alien Abduction: The Inside Story" (*New Scientist*, November 19, 1994, pp. 29–31).

hard-nosed lecturers of psychology, neuroanatomy, and pharmacology. "This clash was an intellectual challenge to me," she said, "and that's really where my enthusiasm for the subject took off."

She was already sympathetic toward the psychic perspective when an out-of-body experience clinched the matter for her. She had been playing with the Ouija board all night with her hirsute psychic pal Kevin and Vicki, another friend. Blackmore was exhausted as a result of too many late nights partying and studying; she had also smoked a little marijuana. Sitting cross-legged on the floor, the room around her suddenly vanished. She was moving down a road lined with trees whose branches met overhead, forming a tunnel. She heard a thunderous galloping noise.

Vicki, as though from a great distance, asked if Blackmore wanted coffee. Blackmore did not, could not, answer. "And Vicki kind of went, 'Oh, well, don't answer then.'" Meanwhile, Kevin, guessing that Blackmore was in a trance of some sort, asked, "Where are you, Sue?" Then the tree-lined road vanished and Blackmore was back in the dormitory room, floating near the ceiling. She saw her body beneath her, still sitting cross-legged on the floor. Her body said, "Oh my God! I'm on the ceiling!" Kevin said, "Wow! It's astral projection!"

She was thrilled to see a whitish strand running from her ethereal self on the ceiling to her material self on the floor. She had read in theosophist texts about the "silver cord" that links the astral self to the body, and here it was! At Kevin's urging, Blackmore flew out of the room, out of Oxford, and out of England. She traveled to the Mediterranean Sea, to New York City, to other places she had always wanted to visit.

Later in the trip, after she had returned to her dormitory room, she felt herself expand, like Alice in Wonderland. She swelled beyond the room, beyond Oxford, beyond the earth. Before long she had become "the entire universe expanding at the speed of light, taking everything in." She felt no distinction between herself and the cosmos; they were one. At the same time, she knew there would always be infinite expanses beyond her. This revelation seemed to come from some benign, powerful presence hovering on the horizon. "It was as though I was arrogantly thinking, 'Ooooo! I've got unified,'" she said. And something rebuked her by saying, "Oh, no you don't. There is always something more."

Her paranormal perceptions continued for several days. She kept drifting out of her body. "Riding my bicycle was difficult, because I would drift off to the side and see myself." She saw auras surrounding other people. "If you were sitting there a day after my experience, I would be able to feel the edge of your aura and put my finger in," she said, reaching her hand toward my chest.

At the time, she didn't know enough about physics or the other sciences to realize how much they contradicted her burgeoning psychic beliefs. "I knew ESP wasn't believed in by all my lecturers at Oxford, but I thought they could be wrong." Her ignorance allowed her to devise theories to explain astral projection, ESP, and other occult arcana. She became entranced by the theosophical concept of "thought forms," which holds that every thought that has passed through a human mind is stored in a kind of eternal psychic ether, which can be tapped into with proper training.

This theory helped her to explain some odd features of her out-of-body experience. While she was soaring over the Mediterranean, she spotted an island shaped like a perfect star — too perfect to be real. The gutters and chimneys of the buildings she flew over in Oxford also differed from their real-life counterparts, she realized later. She decided, initially, that she must have been traveling through the realm of thought forms rather than the real, material world.

Now she views her entire experience as "a very interesting and very particular and quite meaningful hallucination. *But* a hallucination for all that." When people tell her about similar experiences, she does not question their sincerity. "I know memory isn't perfect, and I know they'll adjust the story and everything else. But basically, I simply believe what they are telling me." She nonetheless rejects attempts to give astral projection and other paranormal experiences some sort of objective reality. "You get these people saying, 'Oh, you know, quantum mechanics makes anything possible.' Bullshit!"

Chasing the psi red herring

Blackmore arrived at this skeptical position only after years of investigating paranormal claims. In the psi research community, believers are known as sheep and doubters goats. When her career as a parapsychologist started, Blackmore was a hard-core sheep. She wondered not whether psi works but how. She suspected that psi might be a form of memory recollection rather than direct perception. Reasoning that memories tend to be stored according to concepts and categories rather than their specific visual form, she tested her hypothesis by examining how subjects' guesses go awry in remote-viewing experiments.

In one test, she showed students images of a caterpillar, a butterfly, a train, and other miscellaneous objects and asked them to guess which one she had locked in a drawer. Blackmore had chosen the picture of the caterpillar, and because she believed in psi, she expected the caterpillar to receive the most votes. The point of the experiment was really to see which image came in second. If her memory theory was correct, the butterfly would come in second, because it was conceptually closest to the caterpillar. If the perceptual theory was right, the train would come in second, because the train image resembled the caterpillar image more than the butterfly image did. To Blackmore's dismay, the students' choices were distributed according to chance. Variations of the experiment produced the same results.

When her tests of remote viewing and other psychic phenomena failed to produce positive results, she assumed that she must be doing something wrong. Frustrated, in 1979 she visited the Cambridge laboratory of Carl Sargent, a parapsychologist whose experiments were generating strong evidence of extrasensory perception. Blackmore's impression was that Sargent's experiments were designed well enough to eliminate erroneous results. After observing the researchers for a week, however, Blackmore turned up "a number of errors in the way the protocol was followed," as she put it later. She discovered, for example, that images used in Sargent's remote-viewing experiments were stored in a manner that might allow someone to manipulate the data. When she informed Sargent of her findings, he banned her from his laboratory. The episode left her with grave doubts about other alleged confirmations of psi.

Blackmore also carried out investigations of astral projection, near-death experiences, and other phenomena in which the mind seems to exist independently of the body. She tested whether people who claimed to be able to travel astrally at will could visit her home and identify a number she had taped to a wall. None succeeded. Blackmore eventually concluded that out-of-body experiences are just a particularly dramatic consequence of our cognitive tendency to construct models of ourselves, which we may observe as if from the outside. While dreaming or recollecting memories during ordinary consciousness, we often see ourselves as if from another vantage point.

Blackmore derived a purely physiological explanation for tunnel visions. Such visions are often precipitated by oxygen deprivation or exposure to certain toxins and drugs, she pointed out, factors known to trigger the random firing of neurons in the visual cortex. She and a colleague performed computer simulations showing how the propagation of these random neural discharges through the visual cortex could produce an image of a bright disk expanding against a dark background, which is perceived as a light at the end of a tunnel.

Blackmore also carried out experiments to determine why so many people — one in three, according to surveys in the United States — believe in telepathy and other psychic phenomena. She found that sheep, those who believe in psi and claim to have experienced it themselves, detect patterns in random, Rorschach-type images more readily than nonbelieving goats. Sheep are also more prone to misjudge probabilities. If someone flips a coin and it comes up heads five times in a row, the odds that the next flip will also come up heads are still 50 percent. Sheep tend to underestimate or overestimate the odds because they assume that the previous tosses somehow influence the next one. Similarly, when asked to write down a series of random digits, sheep are more likely to avoid repeating digits or putting two consecutive ones together, when chance dictates that these coincidences should occur quite often. Conversely, given a string of random numbers, sheep overestimate the significance of repetitions.

Of course, all of us are prone to errors like these, Blackmore emphasizes. Asked to guess the number of randomly selected people required to have an even chance that two will share the same birthday (disregarding the year of birth), most of us guess much higher than the correct answer, twenty-two. One reason, Blackmore says, is that we implicitly view such questions in personal terms. We think, How many people would be required for there to be an even chance that someone will have the same birthday as *me?*

The laws of probability can account even for seemingly spectacular coincidences, according to Blackmore. For example, anyone who dreams of the death of an acquaintance or a public figure and discovers the next day that the person just died might be tempted to impute psychic significance to the dream. But a statistical analysis shows that such coincidences are not as unlikely as one might think. The analysis assumes that each of us has only one dream about the death of someone else in our entire lives. Even with that conservative estimate, in a country with the population and mortality rates of the United States, more than a hundred people a year should have dreams of death that "come true" through sheer coincidence. Of course, there are probably many more such dreams, given that we are more likely to dream about the death of someone who we know is ill or at risk in some way.

To the chagrin of her former comrades in the paranormal fold, Blackmore turned from a credulous sheep into a skeptical goat. She told the story of her disillusionment with psi in 1986 in her autobiographical book *In Search of the Light.* She also wrote books and articles offering reductionist explanations of out-of-body and near-death experiences. She did not rule out the possibility of psi; she said merely that she has not verified any claim that she has investigated personally. "Everywhere I have looked psi has seemed to slip away from me. And I cannot look everywhere."

At the end of *In Search of the Light,* Blackmore described psi as a red herring that leads us astray in our search for truth. Parapsychology is often touted by true believers as "the science that will force a new paradigm to topple the old and even serve as a route to spirituality," she wrote. "But it does not deliver." Parapsychology has failed not because psi doesn't exist, but because psi has "nothing to do with the real mystery." The real mystery is not the exotic phenomena that parapsychology addresses — such as extrasensory perception and astral projection — but the ordinary reality and ordinary consciousness in which we are immersed each and every moment of our lives.

Waking from the Meme Dream

In *Beyond the Body,* an investigation of out-of-body experiences originally published in 1982, Blackmore devoted a chapter to lucid dreaming. References to this phenomenon date back at least to Aristotle, and the term *lucid dreaming* was coined by the Dutch psychiatrist Frederick van Eeden in 1913. Just a few decades ago, some neuroscientists still insisted that lucid dreaming is a contradiction in terms; if we know we are dreaming, we must be at least semi-awake. Then in the late 1970s, Keith Hearne of the University of Hull in England and Stephen LaBerge of Stanford independently demonstrated the reality of lucid dreams.* Both employed lucid-dream

---

*Stephen LaBerge, who remains the leading investigator and promoter of lucid dreams, touts them as an all-purpose self-improvement tool that can help us confront our worst fears, tap

adepts who signaled with prearranged eye movements that they were lucid while EEG readings confirmed they were in REM (rapid eye movement) sleep, when dreams occur. (Eye muscles can be controlled during REM sleep, whereas most other muscles are paralyzed.)

Surveys by Blackmore and others indicate that as many as seven in ten people can recall at least one lucid dream, and some people have such dreams regularly. Virtually anyone can learn to have lucid dreams at least occasionally, according to Blackmore. Simply thinking about lucid dreams as you fall asleep will help, as will recalling your dreams and writing them down each morning, or constantly asking yourself "Is this a dream?" while you are awake. Blackmore's skepticism helped trigger her lucid dreams. During her first, she was riding on a ski lift, admiring the view, when she looked down and saw she wasn't wearing skis. "This is daft," she thought — and abruptly realized she was dreaming.

In the 1982 edition of *Beyond the Body,* Blackmore seemed interested in lucid dreams primarily because of the insights they might provide into out-of-body experiences, which have more dramatic paranormal implications. Many people who have lucid dreams also have out-of-body experiences, and vice versa, and lucid dreamers often see their own bodies as if from the outside, but in the 1992 edition of *Beyond the Body,* Blackmore added a postscript that spelled out a profound implication of lucid dreams. "[J]ust as you become 'lucid' in a dream, and realize that it is all illusory," she said, so it is possible to become aware of "the constructed nature" of the self and of all reality even while awake. "In this way you can see into the essential emptiness of it all and the connectedness of everything which can be experienced. It is simultaneously total aloneness and complete oneness. This is a key insight into the mystical experience."

This postscript must have confounded anyone who did not know that in the previous decade Blackmore had become a Zen Buddhist. She was in-

---

our unconscious creative powers, merge with God, or practice a sales pitch. In 1987 LaBerge founded the Lucidity Institute, a for-profit company based in Palo Alto, California, that markets seminars, books, tapes, and gadgets for inducing lucid dreams. One device consists of a mask containing an infrared sensor and a tiny light. When the infrared sensor detects the eye twitches characteristic of REM sleep, the light flashes, ideally making the sleeper lucid without completely waking her up. See LaBerge's books *Lucid Dreaming* (Ballantine Books, New York, 1985) and *Exploring the World of Lucid Dreaming* (cowritten with Howard Rheingold; Ballantine Books, New York, 1990). I profiled LaBerge in "Lucid Dreaming Revisited" (*Omni,* September 1994, p. 44).

troduced to Zen in the early 1980s, just when she was losing faith in psi, by an animal behaviorist named John Crook. Buddhism, he told her, is a form of spirituality that is compatible with science. Both the philosophy and the practice of Buddhism, he argued, can provide clues to understanding consciousness that might complement conventional scientific investigations.

Blackmore became convinced that Buddhism did indeed anticipate some of the latest theories emerging from cognitive science. For example, a Buddhist concept called *anatta,* or "no self," holds that if you eliminate all the passing thoughts, perceptions, and emotions that constitute your mind, you will realize it does not truly exist as a discrete, independent entity. This view echoes the suggestion of some cognitive scientists that there is no unified self at the core of each individual; the self is just a byproduct of the interaction of a host of cognitive functions.

The concept of memes has similar implications. Coined by the evolutionary biologist Richard Dawkins in his 1976 book *The Selfish Gene,* the term *meme* refers to ideas, beliefs, perceptions, and habits passed from one mind to another; memes are to human culture what genes are to human physiology. In her 1999 book *The Meme Machine* (for which Dawkins wrote a foreword), Blackmore argued that what we think of as a conscious self is really a collection of memes; if you take all the memes out of a human mind, you will have nothing left. "The self is just a flowing *story* constructed by memes acquired from other people, and nothing more," Blackmore told me. "It is ever-changing. It is not a thing that can have consciousness and free will."

It is one thing to know intellectually that the self is an illusion and quite another to know it firsthand. Buddhism holds out the promise of experiencing that insight through meditation and other practices. Blackmore participated in her first meditation retreat at John Crook's farmhouse in Wales in 1982, and she started to meditate regularly thereafter. She found it difficult at first. "You know how hard it is. You say, 'Oh, yeah, I'm gonna meditate! Yeah, great!' And then you sit down and meditate for ten minutes and it's, 'Oh, God, I'll do it next week.'" But she stuck with it. She also practiced mindfulness, a Buddhist technique that calls for paying full attention to the present moment rather than being diverted by thoughts of past and future. The technique did not come to her easily. "You can't say, 'I'm going to go on doing it,' because that is being out of the moment. You have to just do it. Now! And now! And now!"

Blackmore now tries to practice mindfulness when her schedule permits. She also meditates every morning, usually for about ten minutes. That is enough to "ground your life in a little bit of openness and peace and letting go," she said. She still has much to learn from Zen — or rather, to unlearn. She emphasizes that Zen aims not at acquiring knowledge, insights, and capacities but at shedding beliefs, obsessions, and other psychological encumbrances. "There's an endless supply of stuff — junk! — in my head to throw away," she said.

Belief in an afterlife was one of the first pieces of junk to go. Some Buddhist teachers have tried to persuade Blackmore that reincarnation is real and that she might recollect her past lives if she were more spiritually advanced. "I sort of think" — Blackmore emitted a noisy Bronx cheer — "it still doesn't make sense." The individual self "is just a bunch of memes sustained by the information-processing system in the brain," she said cheerfully. "When the brain goes, the bunch of memes goes."* Blackmore thought Buddha suggested as much when he told his monks, "'Actions exist, and also their consequences, but the person that acts does not.'"

Blackmore has had flashes of the mystical self-transcendence referred to in Zen as kensho. In fact, she includes her out-of-body experience back at Oxford among them. She views that experience as a hallucination, but a profoundly meaningful one. She has taken to heart the lesson imparted to her toward the end of her journey, that no matter how much we learn and grow, there is "always something more." As a result of that lesson, she views mystical experiences not as ends in themselves but as way stations on a never-ending journey. "One's attitude toward these experiences should be along the lines of, 'Oh, how nice. There's a kensho experience. Keep work-

---

*The neuroscientist and Buddhist Francisco Varela, who agreed with Blackmore that cognitive science is confirming the Buddhist concept of *anatta,* disagreed with her about life after death. When I interviewed Varela in Cambridge, Massachusetts, in December 1999, he told me that during a recent operation for liver cancer he had a near-death experience that "removed any doubts that consciousness is really the most intrinsic part of being." The episode convinced him that death "is not this big leap. It is more like almost a feeling of coming back, coming back home." Varela died eighteen months later. Ironically, in a book of conversations between the Dalai Lama and a number of scientists, edited by Varela, the Dalai Lama expressed doubt that out-of-body and near-death experiences confirm life after death. When the neuroscientist Jerome Engel confessed his fear of death and asked the Dalai Lama point-blank if there is an afterlife, he replied, "That you have to figure out for yourself!" (*Sleeping, Dreaming, and Dying,* Wisdom Publications, Boston, 1997, p. 212.)

ing,'" she said. "If you start clinging to them, then you lapse into, 'Oh goody me, I'm enlightened.'"

On the other hand, Blackmore does believe in enlightenment, which she defines as total self-transcendence, or "waking from the meme dream." She sees enlightenment not as a particular experience or state or insight but as the "loss of a whole lot of crap." You become permanently mindful, in the moment, free from the bonds of the self, even as you continue with the business of living. This is her understanding of the Zen aphorism "Before enlightenment, chop wood, carry water. After enlightenment, chop wood, carry water."

Blackmore has turned her skeptical eye on her own spiritual practices, meditation and mindfulness. One study suggested that the brain waves of yogis in deep meditation remained unchanged even when cymbals were clashed near their ears or their feet were dunked in cold water. Another experiment found, conversely, that Zen meditators do not become habituated to repeated stimuli, as we ordinary folk do. Each time a meditator heard a tone or saw a flashing light, his brain waves responded as if he were perceiving the stimulus for the first time. Studies of other meditators, Blackmore commented, have failed to resolve these contradictions.

Studies concluding that meditation can reduce stress and anxiety, promote healing, and enhance creativity are similarly flawed, she contended. Many of these reports have come from practitioners of Transcendental Meditation. Given the far-fetched claims TMers have made over the years — that they can levitate, reduce crime rates, and alter weather patterns with their focused thoughts — their scientific credibility is particularly shaky, Blackmore said. Some assertions of TM researchers have backfired. After one group contended that the electrical emissions of the brain's two hemispheres become "coherent" — oscillating in accord with each other — during meditation, the British neuroscientist Peter Fenwick pointed out that hemispheric coherence is also a byproduct of comas, epileptic seizures, and death.

Blackmore knows from firsthand experience that mindfulness can have annoying and sometimes dangerous side effects. During one seven-week period when she tried to remain continually mindful, she had difficulty driving. She could not work on page proofs of her writings. Once, walking across a road, she was almost struck by a car. On the other hand, her relationship with her two children — then two and four years old — improved. "Kids live in the present moment, and if you live there with them, it's much more alive." When her kids screamed at her, she screamed back. "And then two seconds later they stop screaming and shouting, and then you stop. You don't have any of this grownup stuff: 'I'm still cross.' It's just gone."

Ideally, Blackmore believes, both mindfulness and meditation help us live within the present. They also help us see what cognitive science has discovered through objective inquiry: We do not perceive the world directly, as it truly is; we actively construct it. We construct ourselves, too. Our ordinary waking self is as artificial, invented, and illusory as the ethereal double selves we hallucinate in dreams and out-of-body experiences. Once we learn this fact and fully realize it in every moment of our lives, we can wake from the meme dream.

Blackmore's view of enlightenment as the "loss of a whole lot of crap" made her suspicious of complicated mystical philosophies, such as that of the transpersonal theorist Ken Wilber. Blackmore had found some worthwhile ideas in Wilber's books, and she respected his attempt to reconcile the enlightenment of the East with that of the West. But his approach to mysticism reminded her too much of theosophy, the ornate, occult philosophy with which she had been enamored when she was still a sheep. Wilber is "one of these guys who turn it all into diagrams and listings," she said. "I don't think he succeeds in bringing together psychology and mysticism in the way that I think he would like to do."

Blackmore was intrigued when I told her about Wilber's claim that he can maintain his mystical awareness even while sleeping. After she first heard about this mystical capacity, she tried to cultivate it herself by staying aware as she fell asleep — "paying attention to the process of sleep and watching the hallucinations begin and so on. But then I'm off, and eight hours later it's morning." Never having met Wilber, Blackmore could not judge whether he is as mystically awakened as he purports to be.

But she had met at least one person, an Australian psychologist named John Wren-Lewis, who she believed might be enlightened. Wren-Lewis was once a self-described "Freud-style skeptic about all things mystical." But after a near-death experience in 1983, he awoke to a mystical awareness that has never faded — or so he claimed. When Blackmore met Wren-Lewis on a trip to Australia, she was immediately "convinced that what he said was quite genuine. And the lovely thing he says is, 'Oh, all this, it's just the universe John-Wren-Lewis-ing.'"

Wren-Lewis contends that enlightenment should not require arduous practices. "You don't have to go anywhere else or do anything in particular. It's right here now," Blackmore said. "Now *he* managed to have that experience and maintain that attitude, awareness, whatever. But most don't," she added. "I don't know what it is about him. Maybe his gentleness." Blackmore cannot remain mindful and selfless all the time, as Wren-Lewis apparently does, but she has her moments. "A lot of the time it's" — scowling, in a snarly, mock-pompous voice — "Sue Blackmore living her life. '*Nnn nnn nnn!* I'm gonna write this book, and everyone is going to buy it, and I'll be so famous!' But it can also be" — soft, singsong voice — "just the universe Sue-Blackmore-ing."

The perpetual-happiness problem

As much as any scientist I have met, Susan Blackmore disproves the decree of the evolutionary biologist Edward O. Wilson that you must choose between science and spirituality, that you cannot be truly committed to both. Blackmore is obviously not religious in any conventional sense. In *The Meme Machine*, she called religions "memeplexes," intellectual viruses that have survived not because they are true but because they excel at replication and infection. To put it another way, religions are just extremely successful chain letters.

Blackmore doubted whether humanity "will ever be entirely free of religion," because natural selection may have made us susceptible to religious beliefs. Normally, she said, natural selection works at the level of individual organisms, but religion might be a rare case of an adaptation that took hold because it benefited groups of individuals. Societies bound by a common religious belief — for which they would fight to the death — might have been more cohesive and hence more likely to survive than societies ruled solely by self-interest.

But natural selection has also endowed us with curiosity, skepticism, and a desire for understanding and gaining control over nature. Science, Blackmore said, is the result of these innate propensities. "Science is not perfect," she emphasized. Scientists make mistakes, commit fraud, and can be faddish, adhering to theories for reasons that are not strictly rational. "False theories can thrive within science as well as within religion, and for many of the same reasons," she remarked. "Comforting ideas are more likely to last than scary ones; ideas that exalt human beings are more popular than those that do not." What makes science superior to religion is that science puts its memes to the test. In fact, science represents not ultimate truth but rather a method for distinguishing true memes from false ones.

Blackmore offers us what even she admits is a "scary" worldview. Our reality, she wrote in *The Meme Machine,* "is just the evolutionary process of genes and memes playing itself endlessly out — and no one watching." She nonetheless promises that we can achieve a kind of ecstatic liberation and joy if we abandon the God meme, the afterlife meme, the free will meme. Blackmore's stance is not free of inconsistencies. She defines spiritual progress as the shedding of beliefs, or memes, that impede our vision of reality. You could argue that Blackmore has merely traded her belief in parapsychology, theosophy, and other occult memes for belief in Darwinism and cognitive science and Buddhism, memes that make some questionable claims of their own.

To demonstrate the illusory nature of the self, Buddhists and certain cognitive scientists — and many postmodern literary theorists, for that matter — ask, Where, exactly, is the self? Of what components and properties, exactly, does a self consist? Buddhists, cognitive scientists, and postmodernists suggest that since no answer to these questions is entirely adequate, the self does not really exist. But all this line of reasoning demonstrates is that the self is a so-called emergent phenomenon that is more than the sum of its parts. The United States cannot be defined in strictly reductionist terms either, but does that mean it doesn't exist?*

Buddhism and cognitive science have also convinced Blackmore that true self-fulfillment and freedom come from accepting that there is no self and no freedom. On the surface, she said, this doctrine might seem to be "a recipe for selfishness and wickedness, for immorality and disaster," but the outcome is just the opposite. When you stop living for your self, "guilt, shame, embarrassment, self-doubt, and fear of failure fade away" and you "become, contrary to expectation, a better neighbor." In other words, true self-abandonment leads not to the moral nihilism and sociopathy exhibited by some gurus, but to the gentleness of John Wren-Lewis, the Australian psychologist who Blackmore believes may be totally enlightened.

A closer look at the case of Wren-Lewis raises various questions

---

*One mystical expert who looks askance at the promotion of the Buddhist no-self doctrine by Western scientists is Ken Wilber. The no-self concept, Wilber told me, is just an idiosyncratic view of a single branch of Buddhism. "It lends itself, unfortunately, to a scientific, materialist reductionism," he explained, which holds that "there is nothing but this mindless materialism, shuffling on down the road."

about enlightenment. In an article in the spiritual journal *What Is Enlightenment?*, he recalled that he was traveling in Thailand in 1983 when he was poisoned by a would-be thief and collapsed.* When he awakened in a hospital, Wren-Lewis's sense of himself as a discrete entity separate from the world had vanished. He felt as though a "radiant dark pure consciousness" was pouring into the back of his head. The sensation was so strong that Wren-Lewis touched the back of his skull, "half wondering if the doctors had sawn part of it away to open my head to infinity." He kept waiting for this "joy beyond joy" to dissipate, but it never did. He no longer feared death, and physical pain became "simply an interesting sensation, another of nature's wonders." Like Ken Wilber, he maintained his awareness even while he was asleep.

Reading about Wren-Lewis's preternatural cheerfulness reminded me of an episode of the television comedy *Ally McBeal.* The main character was a businessman transformed by a benign brain tumor from a scowling, cutthroat Scrooge into a perpetually jolly, Santa-like figure. Eager for others to share his happiness, he increased the salaries and benefits of his employees. When his partners took him to court to prove he was incompetent, so they could wrest control of the company from him, the businessman argued that his behavior was quite rational. After all, what could be irrational about being happy and wanting others to be happy? Then the businessman's wife died. To his horror, he felt no sorrow, though she was the only person his old self had cared for. He couldn't even wipe the smile off his face. He had the tumor removed, and he became a miserable, nasty Scrooge once again, but a Scrooge who could grieve for the one person he ever loved.

**What Is Enlightenment?* is published twice a year by Moksha, an organization based in Lenox, Massachusetts, and founded by the guru Andrew Cohen. The magazine serves as a promotional vehicle for Cohen, who claims to have become enlightened, or "awakened," in 1986. It contains articles by him and advertisements for his books, videos, and retreats. Photographs show him striking the classic guru poses, laughing sublimely or gazing heroically into space. But the magazine also features articles by many other spiritual authorities, some with views that contradict Cohen's. Each issue wrestles with a different topic: the tension between science and mysticism, the Westernization of Eastern religions, the commercialization of spirituality, and the relation between sexual and spiritual liberation. One of Cohen's first devotees was his mother, Luna Tarlo. She told the story of his ascent to guruhood and her eventual disenchantment with him in *Mother of God* (Plover Press, Brooklyn, New York, 1997).

If we become immune to ordinary grief and pain, are we still fully human? That is the puzzle posed not only by the fictional businessman but also by the case of John Wren-Lewis. If we decide to ignore this philosophical quandary and pursue enlightenment anyway, the next question is, What is the best way to achieve it? According to Wren-Lewis, there is no way. Enlightenment, he says, consists of existing wholly in the present and hence transcending time. Wren-Lewis would therefore look askance at the assertion of Ken Wilber and other transpersonal philosophers that enlightenment represents the culmination of an evolutionary or developmental process.

In fact, Wren-Lewis wrote, "the very impulse to seek the joy of eternity is a Catch-22, because seeking itself implies a preoccupation with time, which is precisely what drives eternity out of awareness." Zen and other spiritual traditions, he speculated, far from helping us become enlightened, may ensure that enlightenment will always elude us. Wren-Lewis thought that enlightenment might come about only through sheer happenstance, or what Christians sometimes call grace. Of course, Wren-Lewis's own transformation, like that of the businessman on *Ally McBeal,* almost certainly resulted not from grace but from brain damage that he sustained when he was poisoned.

Hyperventilation and Derealization

Although Blackmore and Wren-Lewis may disagree about the efficacy of Zen and other spiritual practices, they concur that total mindfulness, self-transcendence, and waking from the meme dream should engender joy in us. But both Wren-Lewis's blissfulness and Blackmore's more conventional cheerfulness may be independent of their mystical self-transcendence. There is no guarantee that when we perceive the illusory nature of ourselves and of ordinary reality the world will appear "glorious," as Blackmore put it.

Sensations of unreality are quite common, enough so that psychiatrists have given them labels: derealization and depersonalization. In the former, you become alienated from external reality; everything seems unfamiliar, strange. In the latter, you become a stranger to yourself as well; you become detached from your own emotions and perceptions. Brain scans have tentatively linked depersonalization and derealization to excitation in neural regions called the temporal gyrus and mesolimbic structures. When these sensations "cause marked distress or impairment in functioning," declares the *Diagnostic and Statistical Manual of Mental Disorders,* the bible of modern psychiatry, they should be considered pathological and deserving of treatment.

In retrospect, some of my earliest forays into altered states gave me a taste of depersonalization and derealization. When I was in grade school, my pals and I engaged in a practice that might be called assisted hyperventilation. One of us — Tommy, let's say — would huff and puff a half-dozen times before sucking in a final, voluminous breath and holding it. At that instant, someone would wrap his arms around Tommy's chest and squeeze. Tommy would faint and fall in a heap on the floor, twitching and grunting in comical fashion for thirty seconds or so while the rest of us stood over him, laughing.

I watched several friends undergo the procedure — and suspected that they were faking their faints — before I tried it myself. The experience was much more wrenching than I expected. One moment I was firmly embedded in the quotidian reality of a mid-twentieth-century preadolescent American boy goofing around with his buddies. The next moment I was looking up from the bottom of a well at a pack of goblins jabbing their fingers at me and shrieking gleefully. Only gradually did I realize that these gibbering demons were my friends, Tommy and Chris and Mike. Only gradually did I remember my own name and identity.

The third or fourth time I was knocked out, my sense of unreality and alienation did not quickly dissipate after I regained consciousness. My rational cognitive machinery put name tags on things, and I resumed functioning more or less normally. But viscerally, emotionally, I remained estranged from my friends, from myself, from the world into which I had been thrust. The feeling persisted for weeks, forming an ominous, shadowy backdrop for all my social interactions.

After my hallucinogenic trip in 1981, I experienced more severe episodes of depersonalization and derealization. I would be in some mundane situation — ordering a ham sandwich in a Manhattan deli, sitting in a classroom listening to a professor pontificate about *Finnegans Wake* — when sensations of strangeness and unreality would overcome me. The world around me wavered and trembled like a membrane on which images were being projected. I was terrified that at any moment the membrane would burst and the blackness behind it would come flooding through and annihilate not only me but the entire world. My intellect told me: This is dumb, this is delusional; I am projecting my own fear of insanity and death onto the world. But emotionally, I was convinced that only my own willpower was preventing the end of everything.

I also had an eerily visceral sense of what Hindus call Thou Art That, of my mystical identity with others. When I made eye contact with friends or strangers on the street, an electric current seemed to flow out of my eyes and into theirs, down through their bodies and feet and up through my feet and body to my eyes again, completing the circuit. The paradoxical effect of this sensation was to alienate me from others. I shunned eye contact, and contact in general. Fortunately, these flashbacks faded after a few months.

According to a Tibetan Buddhist practice called dream yoga, cultivation of lucidity during dreams helps you realize that waking reality, too, is just a dream. Dream yoga also supposedly helps you stay calm during the *bardo,* the terrifying aftermath of death, so that you can safely navigate to nirvana, or into another incarnation, if you prefer. The ancient Hindu texts the Upanishads claim that the only reality is the formless, infinite, eternal void from which all things come and to which they return. All else, including your mortal self, is unreal. Doctrines such as this give us a perverse sort of immortality: We cannot die, because we do not exist in the first place. The Upanishads promise that when we know, really *know,* this fundamental truth, we will achieve nirvana, bliss, heaven. To me, seeing the world and myself as unreal felt more like hell. Those who are enlightened — blissfully enlightened — must somehow sidestep or push past this dreadful state. But how?

7

ZEN AND JAMES AUSTIN'S BRAIN

AN EIGHT-YEAR-OLD BOY named James is strolling on a city sidewalk with his mother. As he looks down a side street he is startled to see himself reflected in several storefront windows. This vision of images of himself outside himself triggers an identity crisis in the boy and an acute sense of unreality. He wonders, "How really real is our ordinary life? How does one know which reality is the real one?"

The boy was experiencing the psychological syndrome called derealization. Derealization and its twin syndrome, depersonalization, are common side effects of mental illness, mind-altering drugs, and lack of sleep, but they also occur spontaneously in healthy people. Surveys of high school and college students have found that up to 46 percent have undergone brief episodes of depersonalization. Meditation often induces depersonalization and derealization, as it undermines your conventional sense of self and reality. Ideally, however, estrangement yields to bliss as you become immersed in a transcendent, vastly expanded self and reality. Enlightenment, the supreme form of mystical awareness, can be described as *neo*realization and *neo*personalization.

I found this discussion of depersonalization and derealization and their relation to enlightenment in James Austin's *Zen and the Brain.* Austin is a Zen Buddhist and a neurologist, a specialist in disorders of the brain. The boy disoriented by the sight of his own reflections was Austin himself. This combination of the personal and the scientific, the objective and the subjective, typifies Austin's approach to mysticism. *Zen and the Brain* interweaves scientific facts and hypotheses about the brain, mind, and mysticism with stories from Austin's life, particularly from his experiences as a student of Zen.

Unlike Andrew Newberg, Michael Persinger, and Susan Blackmore, Austin has conducted little mysticism-related research on others; his primary research subject is himself. But his book represents one of the most thorough, not to say exhaustive, reviews of the neurophysiology of mysticism to date. Austin professes no interest in defending the perennial philosophy — or any other philosophy, for that matter. He calls his outlook "perennial psychophysiology," to emphasize its grounding not in philosophy or theology, which he views with suspicion, but in brain science.

Austin sees mysticism as a route to profound changes in personality rather than to profound metaphysical insights. You become calmer, more stable, less fearful, less selfish, more compassionate; you achieve a kind of transcendent sanity and maturity. Hence Austin works especially hard to distinguish healthy mysticism from depersonalization, derealization, and other psychopathologies. Neurologists, he notes, have learned a great deal about how brains function by examining brain-damaged individuals. In the same way, scientists may glean clues about the neurophysiology of mysticism by studying disorders that mimic or diverge from mystical awareness in specific ways.

Schizophrenics, he says, occasionally experience something like mystical bliss. Some plunge into blackout states resembling the meditative state known as absorption, in which you lose all contact with external reality. But compared to schizophrenics, healthy mystics retain more control of their thought processes and behavior; they do not withdraw from social interaction as much, and they rarely have the auditory hallucinations that characterize schizophrenia. Schizophrenics and mystics both have a heightened sense of meaningfulness. But whereas a schizophrenic sees everything as a personal message intended for him alone — the scandal of particularity! — the healthy mystic discerns meaning of a more impersonal, general nature.

Other brain disorders, Austin points out, produce effects resembling negative images of mystical experiences. Apraxia undercuts the ability to initiate complex actions. In contrast, mystical awakening gives you a "promethean hyperpraxia," an ability to translate intention into action more rapidly, effortlessly, and creatively. People who suffer from prosopagnosia cannot consciously recognize faces, even though lie detectors indicate that they are responding to the faces on an unconscious, "gut" level. In kensho, Austin commented, the reverse may be true; that is, you retain your ability to recognize what you are seeing, but you no longer respond to it on a personal, emotional level. Then there is a bizarre disorder called simultagnosia, in which you can perceive objects in a scene only individually, not as parts of a whole. Zen, in contrast, boosts the ability to see holistically.

Austin casts doubt on the suggestion — made by Michael Persinger, inventor of the God machine, among others — that mystical states stem from seizurelike excitation in the temporal lobes. Few epileptics experience religious or mystical visions during their seizures, Austin points out, and few mystics show signs of epilepsy. The subjective effects that accompany epileptic seizures differ from those reported by healthy mystics; nor are healthy mystics humorless, rigidly moralistic, and irritable, as religiously obsessed epileptics often are.

Austin shoots down other popular hypotheses about mysticism. Because the human language capacity is based primarily in the brain's left hemisphere, some neuroscientists have concluded that mystical experiences, which supposedly transcend language, must be primarily right-brain phenomena. This thesis is almost certainly too simplistic, Austin asserts; saying one hemisphere is more important for sustaining a given state of mind is like saying one wing of a hummingbird is more important to its flight.

Like Blackmore, the mystical expert whom he most resembles, Austin sees spiritual practice as a process of cognitive subtraction rather than addition. He goes much further than Blackmore in specifying how this process might work on a neural level. He compares meditation and other spiritual practices to sculpture or etching, which also create by taking away. He means this analogy literally. He cites evidence from animal studies that neurochemicals such as L-glutamate, aspartate, and nitric oxide can act as "excitotoxins," which destroy specific types of neurons in specific regions of the brain by overstimulating them. Released by mystical experiences, excitotoxins "can be potent agents," Austin wrote in *Zen and the Brain,* "prompting a kind of highly localized 'etching' away within certain vital regions." As we lose neurons, we also shed beliefs, obsessions, and emotions that distort our view of reality.

A major target of this etching process, Austin speculates, is the cortex, the sheath of neural tissue capping the brain. The cortex imposes self-oriented concepts and labels on incoming perceptions, asking, in effect, How does this affect *me?* Meditation and mystical states may also reduce the dominance of the limbic system, a collection of neural structures deep in the brain that regulates emotion. The limbic system tags incoming perceptions with emotional significance, telling us, This is good, This is bad, or simply, This is important. As the roles of the cortex and limbic system diminish, Austin proposes, so do our innate self-centeredness, narcissism, and anxiety.

But an underlying theme of *Zen and the Brain* is that scientists have established few facts about mysticism — or indeed any mental phenomena — beyond a reasonable doubt. After scrutinizing the scientific literature on meditation, for example, Austin concluded that most studies have been flawed, lacking in controls, carried out in highly artificial settings, with inadequate instruments, and often by biased researchers and subjects. At best, the reports suggest merely that meditation helps many people feel more relaxed, but simply sitting quietly or napping may produce equivalent benefits. In one investigation cited by Austin, experienced practitioners of Transcendental Meditation were found to be sleeping almost half the time they were supposedly meditating.

Austin is so forthright in his review of mysticism that he seems to take two steps back for every step forward. His negative capability abandons him only when he addresses the ultimate goal of Zen and all spiritual practices: permanent enlightenment, total self-transcendence. Austin has no doubt that such a state, which he calls "Ultimate Pure Being," exists. After struggling to describe this condition in *Zen and the Brain,* he concluded that the best way to represent it is not to represent it. To emphasize his point, he left the next page blank. It was not even numbered.

Austin's private Idaho

After spending most of his career at the University of Colorado Health Sciences Center in Denver, Austin retired in 1992 and moved with his wife, Judy, to Moscow, Idaho. Although he turned seventy-five in 2000, he still teaches at the University of Idaho, which is based in Moscow. To reach him, I had to fly to Seattle and catch a prop plane to a tiny airport near Moscow. The driving directions Austin mailed me, complete with a map and precise distances between landmarks, took me through fields of rolling farmland and onto a dirt road. After passing several decrepit barns and a sign saying "Peacock Crossing," I spotted Austin's brown ranch-style house. It perched on the side of a ridge where fields of sprouting wheat and peas gave way to conifers.

Wearing jeans and an old flannel shirt, Austin looked more like a rancher than a Harvard-trained neurologist. He was soft-spoken with a weathered face and rugged, thick-fingered hands. Although the morning was chilly, he recommended that we talk outside. He mentioned that he has been an outdoorsman since his youth, when he spent his summers on a relative's farm in Ohio. "I don't put God with a capital G right up at the top of everything," he said. "I'm pretty much oriented toward Mother Nature in all her guises."

We sat side by side on his front porch, facing not each other but the panorama sprawling before us. Austin identified notable features of the landscape. Through the valley below ran the Clearwater River, which Meriwether Lewis and William Clark navigated two centuries ago in their search for a northwest passage. Beyond the river loomed a dark line of peaks called Paradise Ridge. Through Austin's binoculars, I could just make out the Wallowa Mountains more than a hundred miles away, across the Oregon border. The dim, snowy abutment hovered above the horizon "like the dream of nirvana" — or so I effused in my notebook.

Austin was raised as a Unitarian in Ohio, but his belief in Jesus' divinity and other Christian dogmas faded while he was earning his degree at Harvard Medical School. In 1974, after teaching for seven years at the University of Colorado, he went to Kyoto, Japan, for a sabbatical. On the overseas flight he read a book a friend had given him: *Zen and the Art of Archery.* Intrigued, Austin enrolled for instruction at a Zen center in Kyoto, and he was soon spending three days a week there.

During one meditation session he lost contact with the outside world and with his own self. "It was the most intense blackness imaginable," he said. "And yet it glistened, like being in an obsidian crystal." This experience, which Austin calls absorption, was the seed from which *Zen and the Brain* would grow. "I realized that this path or training was a window into an area of neurobiology that I didn't know anything about," he said. "I also became aware of the power of meditative training to access these extraordinary states of consciousness."

The bliss and enchantment Austin felt during absorption reminded him of the effects of morphine, which he once received for a minor surgical procedure. (His pleasure was so intense that he vowed never to take an opioid again.) Austin hypothesizes that absorption may be accompanied by a pulse of our own endogenous opioids, the best known of which are called endorphins. These compounds may give rise not only to the subjective sensations of bliss, passivity, and fearlessness reported by those who achieve deep meditative trances but also to their slowed respiration and heartbeat.*

Austin suspects that what he calls absorption is what the neurotheologian Andrew Newberg calls absolute unitary being. While he applauds all attempts to unravel the physiology of mystical states, Austin believes the mystical model proposed by Newberg does not do justice to the diversity and complexity of mystical experiences. He also doubts whether the Buddhists and nuns whose brains Newberg has scanned are actually achieving deep mystical states in the laboratory. (Newberg conceded as much when I interviewed him.) Absorption is quite rare, Austin said, and rarer still is what Zen masters call kensho.

Austin experienced kensho in 1982 while he was on a sabbatical in London. It was Sunday morning, and he was standing on the platform of a tube station waiting for a train that would take him to his Zen session. Abruptly, his ordinary sense of self dissolved. Nothing changed, and everything changed. Austin wasn't just seeing; he was *seeing*. The dingy station was still a dingy station, he recalled in *Zen and the Brain*, but it was also "profound, implicit, perfect reality." He understood "at depths far beyond simple knowledge" that "this is the eternal state of affairs." There is "nothing more to do" and "nothing whatsoever to fear," since the world is "completely and intrinsically valid." Austin felt neither ecstasy nor any other emotion, only the "cool, clinical detachment of a mirror as it witnesses a landscape bathed in moonlight." He understood why Zen painters use the cold, pale moon to depict enlightenment.

Austin's kensho had a deeper effect on him than his absorption experience. If absorption is being "rinsed in a cloudburst of warm rain," kensho

---

*At least one study has undercut Austin's hypothesis about the role of endogenous opioids in meditative states, according to Andrew Newberg. The study found that a male meditator was unaffected when given naloxone, a drug that blocks the effects of opioids. When I asked Newberg for a reference, he cited a paper by two researchers in the Department of Pharmacology of the National University of Singapore: "The Effects of Centrally Acting Drugs on the EEG Correlates of Meditation," by M. K. Sim and W. F. Tsoi (*Biofeedback and Self-Regulation*, vol. 17, no. 3, 1992, pp. 215–220). Of course, it is possible and even likely that this meditator simply did not achieve the state of absorption that Austin said releases endorphins.

is being "washed away by a flash flood of clear, cool understanding." During kensho, the external world does not disappear, as it does during absorption, but your ordinary sense of yourself as an individual distinct from the rest of the world vanishes. You no longer view the world "as something to be manipulated," Austin said. "You're just perceiving it as the way things really are."

Kensho transformed Austin. "You lose — really lose — the psychic, intellectual, emotional, visceral sense of your egocentric self," he said. "That coincides with a complete lack of all the visceral ingredients that must underlie our sense of fear." He has felt calmer, more emotionally stable, ever since his kensho, and he no longer fears death as he once did. "Why should I be afraid of dying? I wasn't afraid of being born." He also lost interest in philosophy, theology, and other metaphysical endeavors. "I view a lot of the theological interpretations and all of the philosophical speculations as encumbrances, cultural baggage that needs to be worked through and gone beyond as soon as possible." That is why Austin looks askance at the perennial philosophy. He agrees with perennialists such as Houston Smith and Ken Wilber that "over many cultures for many centuries there have been a rather restricted number of experiences that I would call experiences of insight-wisdom." But each episode of insight-wisdom is unlike all others, even for the same person. Zen is not about dogma and metaphysics; it holds that "the world, just as it is, is immensely satisfying. No words are being imposed between you and the world as it is right now."

Isn't it ironic that he should say such things, I asked, given that he is the author of an intensely analytical, intellectual, and very long book? Austin nodded and smiled at my question. One side effect of practicing Zen is that "one becomes comfortable with these paradoxes," he said. "They lose their polarity." He cocked his ear. "Listen to the geese!" he said. Scanning the valley, he located a wedge of dark forms moving against the pale green farmland. He guessed that the Canada geese were headed for a small pond at the bottom of his property. Sure enough, with much braying and splashing the geese skidded down onto the pond.

We humans, he continued, are compelled to concoct theories, explanations, cause-and-effect relationships, even for those phenomena that are beyond explanation. "That's our glory and our curse." Like Susan Blackmore, Austin has no use for reincarnation or other theories about life after death, and he called the concept of God "unnecessary for my living or loving or being comfortable." (*God* is not listed in the index of *Zen and the Brain.*) Although he is amazed by the marvels of nature surrounding him, he feels no need to attribute them to a divine creator. "We can decide that this is God-inspired. Or we can decide that this is some kind of happening that is *beyond belief.*"

Nor does Austin feel any commonality with mystics who insist that mind rather than matter is the basis of reality. "There isn't any experience I've ever had that would allow me to reach that conclusion," he said. "So to me these arguments are specious. They are distractions that stand in the way of our appreciating clouds." He lifted his hands toward the sky. "I mean, here we are talking about how somebody misuses and attaches meaning to the word *mind.* Big deal. So what. Let them get caught up in all that language. I'm not there."

Austin has no interest in theodicy, the attempt to explain how a benevolent, omnipotent God could allow so much evil and pain in the world. "I think the theologians have made a mess of the human spiritual quest," he said. "They have made it too cumbersome, too unbelievable." This is true of Buddhism as well as other religions, but Buddha in his original teachings did not fret over these sorts of questions. "That's not the Buddha's message. His message was to cut through all this stuff."

I confessed to Austin that I couldn't help fretting over unanswerable metaphysical questions, such as why there is such an unfair allotment of suffering in the world. "I can't answer that personal question for you," he replied. "I think you'll have to answer that question within yourself in the next few decades." Austin was forced to confront life's unfairness while still in medical school, when he specialized in pediatric illness; there is nothing more unfair, after all, than a child with a terminal disease. All he could do, he decided, was "try to adhere to the good, mind my own business, and take care of the suffering of other people." Although "life isn't fair, this is what's on the plate in front of me." His training in biology also helped him to see suffering as a biological fact rather than a theological problem. "As I got farther into biology, it became more clear that nature *was* red in tooth and claw."

Austin never had another kensho like the one in 1982. He suspects that reaching "the stage of ongoing enlightenment" — total, permanent self-transcendence — requires multiple kenshos and complete integration of that mode of perception into one's daily life. Austin doubts whether anyone he has known — even his first Zen teacher in Japan, perhaps the wisest man of Austin's acquaintance — has attained this level of Buddha-like perfection.

But he believes there may be a few rare souls on earth who have achieved full enlightenment. "These people are not going to be out in the open, leading seminars on consciousness, publishing lots of books, or starting centers down in southern California," he said. "They are going to be private people." Austin sensed an unspoken question hanging in the air. "Just to put the record straight, I don't consider myself anything other than a student of the Zen meditative path, whose eyes have been opened by a series of experiences and by contact with some inspiring people."

Looking for the Peacock

Austin's description of spiritual progress as a process of neural, cognitive, and emotional subtraction makes me uneasy. It reminds me of "The Last Hippie," a case history told by the neurologist Oliver Sacks. The protagonist of Sacks's tale was Greg F., a quintessential child of the sixties. As a teenager, Greg left his home in Queens, New York, to live in downtown Manhattan, where he listened to the Grateful Dead and dropped LSD. In search of a more stable spiritual life, he joined the International Society for Krishna Consciousness, one of many Hindu-based groups attracting spiritually inclined American youths. He shaved off his long hair, donned the cult's distinctive saffron robe, and moved into an ashram in New Orleans, where he and his comrades chanted "Hare Krishna" and other mantras intended to trigger enlightenment.

In the early 1970s, Greg's eyesight started dimming, and he became increasingly forgetful. When Greg complained to his guru, he assured Greg that these were signs of Greg's growing spiritual illumination; the swami told Greg's parents the same thing when they tried to contact their son. When the parents finally insisted on seeing Greg, they found to their horror that he was totally blind, hairless, fat, and disoriented, so much so that he did not even realize his plight. He wore what his father called a "stupid" smile and made "idiotic," jokey remarks. Greg's parents rushed him to a hospital, where x-rays revealed an orange-sized tumor in his brain.

After the tumor was removed, Greg required permanent institutionalization. When Oliver Sacks met him in 1977, Greg was "bland, placid, emptied of all feeling — it was this unnatural serenity that his Krishna brethren had perceived, apparently, as 'bliss.'" Ordinarily the gentlest of writers, Sacks added bitterly that Greg's fate "could have been prevented entirely had his first complaints of dimming vision been heeded, and had medical sense, and even common sense, been allowed to judge his state."

I have no doubt that Sacks's fellow neurologist James Austin would have diagnosed Greg's condition at its onset. Austin would never have mistaken Greg's "unnatural serenity" for enlightenment, which Austin defines as "promethean hyperpraxia," the opposite of passivity. I nonetheless wonder whether his neural-etching theory might lead naive mystical enthusiasts to confuse enlightenment with a state that is "placid, emptied of all feeling." Austin does, after all, suggest that Zen training quells our passions; he even depicted enlightenment with a blank page.

If I look hard enough, I can find other bones to pick with Austin. He looks askance at metaphysics, and yet his description of his kensho on the London subway platform — "this is the eternal state of affairs," and so on — was nothing if not metaphysical.* He indulges in special pleading for Zen, which supposedly offers a time-tested, self-critical tradition, unlike "unseasoned 'new-age' mysticism." Zen can be cold, amoral, self-absorbed, doctrinaire, Austin acknowledged, "but it redeems its weaknesses in the act of becoming its own severest critic."

Actually, Zen lore highlights, in an almost fetishistic manner, the sadism and masochism of its sages. The Zen patriarch Bodhidharma is said to have meditated for so long that his legs became gangrenous and had to be amputated. He became so enraged with himself for falling asleep that he tore his eyelids off; this was the origin of the Zen technique of open-eyed meditation. Bodhidharma kept would-be students waiting outside his monastery before he admitted them, to weed out the less dedicated. One supplicant, to demonstrate his seriousness, chopped off his own hand. He

---

*This was the chief complaint made against *Zen and the Brain* in "Is There a Voice in the Night?," by Arthur Deikman (*Times Literary Supplement,* August 6, 1999, p. 30). Deikman, a psychiatrist whose mystical term *deautomatization* is discussed below, said: "After [Austin's] stern warnings against giving metaphysical interpretations to the effects of neurotransmitters and endogenous opioids, he then speaks of his *kensho* — 'awakening' — as an experience of 'things as they really are' . . . But previously he had taken great pains to show us the Wizard of Oz concealed behind the curtain, and the question remains: Is *kensho* anything more than a neurophysiological illusion that we are perceiving the world as it is? If it isn't an illusion, why not?"

went on to become a great master in his own right. In another Zen legend, a teacher slammed his door on the leg of a student trying to enter his room. The student became enlightened at the moment his leg broke.

Austin implies that his cool, hyperrational, relatively anxiety-free awareness is the mystical rule when in fact it may be the exception. His emotion-generating limbic system may play a diminished role in his awareness, but that may not be the case with most other mystics. In fact, other researchers have suggested that the limbic system plays an *enhanced* role in mystical perception. Kensho and other mystical experiences, Austin says, counter our innate selfishness; just as the Copernican revolution shattered our earth-centered view of the cosmos, so does mysticism overthrow our self-centered perspective. But as the careers of some of the professional mystics known as gurus have demonstrated, mysticism is often associated with pathological narcissism.* In other words, Austin's book might have been more accurately titled *Zen and James Austin's Brain*.

Austin grants that his equanimity may not stem entirely, or even in part, from his Zen practice. Aging also prunes the brain of many neurons, and often "your hard edges get knocked off as you get older," he said. In addition, he might have become inured to death and suffering because of his decades of work as a physician. When he compares himself with other physicians with similar backgrounds, they seem more anxious, in thrall to their desires, fearful of death. But his innate temperament, he said, rather than his Zen practice and mystical experiences, might explain these differences, too.

The best way to measure the benefits of spiritual practices such as Zen, Austin suggested, would be to conduct studies of identical twins, one who engages in a particular practice and one who does not. Much could be learned both while the twins are alive, through neuroimaging and other tests, and after they die, through detailed examinations of their brains. Such studies would be extremely difficult and time-consuming, but done

---

*The British psychiatrist Anthony Storr reached this conclusion in his 1996 book *Feet of Clay* (Free Press, New York, 1996), which examined twentieth-century gurus such as Gurdjieff, Bagwan Shree Rajneesh, and Jim Jones, head of the ill-fated Jonestown cult. Storr noted that some gurus have professed clearly delusional beliefs, perhaps triggered by manic depression, schizophrenia, or other mental illness. But what really distinguishes gurus from more orthodox spiritual leaders, Storr concluded, "is not their manic-depressive mood swings, not their thought disorders, not their delusional beliefs, not their hallucinatory visions, not their mystical states of ecstasy; it is their narcissism" (p. 210). Instead of shrinking to a point and vanishing, the mystic's ego may expand to infinity.

properly, they could reveal the relative contributions of nature and nurture to the qualities attributed to sages and saints.*

This is one of Austin's most admirable qualities, his insistence that spiritual suppositions should be tested and not taken on faith. Several times during our interview he told me that he ascribes to the slogan of Missouri: "Show me." I must admit that Austin's presence and manner — his engagement with the world — also impressed me. After we had talked for several hours, he gave me a tour of his land. As Red, his golden retriever, bounded around us, Austin squatted here and there to fondle various flora and inform me of their properties. This fungus is delicious sautéed, that grass is fire-resistant. He is not an indiscriminate nature lover; he plucked out dandelions and a grass that choked out other growth.

The evergreen forest behind Austin's house looked untamed at first glance, but closer inspection revealed evidence of his handiwork. Austin had thrown logs across a brook to create a watering hole for animals. As we stood beside the pool, he pointed to saucer-sized divots in the pool's mud floor: moose. On a nearby hillside, Austin noticed a smaller set of cloven prints. Too big for deer, too small for moose. Must be elk.

Spotting a pile of tarry excrement, Austin genuflected before it, poked it with a stick, noted the fur strands and bone fragments in the broken turds. Given the size and color, probably bear scat, he surmised, although it might have come from a coyote who had enjoyed an especially large and bloody meal. We arrived at a pipe poking out of the ground, a ceramic mug beside it. Water from an underground spring bubbled from the pipe. After letting Red lap the water from his cupped hand, Austin filled the mug and handed it to me. The water was cold and sweet, just as Austin promised.

For an indulgent moment, I wondered if Austin, in spite of his de-

---

*Evidence of a strong genetic component to spirituality, Austin speculated in a startling passage in *Zen and the Brain* (pp. 689–690), might persuade religions such as Tibetan Buddhism to breed new leaders instead of "searching all over for the right baby." A nun and a monk "selected on the basis of their outstanding lineage and capacity" would supply the ovum and sperm for a test-tube baby brought to term in a surrogate mother. "It is not necessary for anyone today to buy into this futuristic scenario or to feel comfortable about it," Austin said. "Our task is to take a deep breath and to accept that such a blessed event is now *technically* possible." Wilder eugenic scenarios are imaginable. Sages like the Dalai Lama could be cloned. If researchers identify specific genes associated with mystical aptitude — call them "mysticism genes" — they could be inserted into fetuses along with the genes for high IQ, perfect pitch, and tennis talent. I critique these "designer-baby" fantasies in *The Undiscovered Mind* (Free Press, New York, 1999, pp. 137–164).

murral, might be one of the rare "living Buddhas" that he suggests are scattered around the earth. But no, I decided, he is just a man at peace with himself and with life — an accomplishment rare in its own right. I would certainly settle for that. Austin cautioned in *Zen and the Brain* that "most aspirants, perhaps authors in particular, will remain at best partially enlightened." Those who live by the word, he seemed to be suggesting, pay a price for it.

The hills were ruddy with the light of the late afternoon sun as I climbed into my car. Austin warned me to watch out for his neighbor's peacock, which likes to strut about in the middle of the road. As my car approached the "Peacock Crossing" sign once again, I braked to a crawl, hoping to catch a glimpse of the improbable beast. It never appeared.

The explanatory chasm

The most common cavil aimed at Austin is that his book is too long — "far too long," Susan Blackmore said in a review, although she also praised Austin for laying the foundations of an "experiential neurology" that "includes insight and wisdom among its aims." Austin did seem to ignore the message of one of his own quotations: "Knowledge is a process of piling up facts; wisdom lies in their simplification." But in its very prolixity, it seems to me, *Zen and the Brain* achieves an uncommon honesty. Austin implicitly conceded that wisdom and simplicity are not yet possible. Scientists studying mysticism are still in the fact-accumulation stage, and may always be. To present a sleek, pared-down hypothesis, mentioning only those facts or pseudofacts that support it (as so many books of this ilk do), would be dishonest.

Austin concluded his book with the obligatory burst of optimism about science's progress in understanding the brain and its effluents: "Vast numbers of leads are converging. They promise to clarify what was only yesterday that black box of a brain." Just as computers and satellites have made long-term weather forecasting possible, so will advances in neuroscience allow us to "study and predict some events of our own 'inner weather.'" But he also warned that there will be no "Rosetta stone" that translates the "subtle coded language of the brain" — the neural code — into "direct experience of extraordinary states of consciousness."

The attempts of scientists such as Austin, Blackmore, Persinger, and

Newberg to understand mysticism in neurological terms evoke a philosophical concept called the explanatory gap. This term, coined by the philosopher Joseph Levine in 1983, refers to the disconnect between physiological theories of the mind and the subjective sensations those theories purport to explain. On the one hand, you have a model consisting of physical objects: neurotransmitters, receptors, synapses, dendrites, ganglia, amygdalas, posterior superior parietal lobes. On the other hand, you have what philosophers sometimes call qualia, the purely subjective perceptions, thoughts, emotions, and memories that make up a mind.

The fact is, neuroscientists cannot explain how the brain carries out the most elementary acts of cognition — for example, how I know the person lying beside me when I wake each morning is my wife. Some prominent scientists and philosophers have reluctantly predicted that the explanatory gap will never be closed. Even if neuroscientists crack the neural code, so that they can determine precisely which neural events are correlated with a given set of mental events, there may always be a strange incongruity between physiological and mental phenomena; something about the mind makes it peculiarly resistant to scientific reductionism. This philosophical position is called mysterianism. You don't have to be a mysterian to wonder whether the explanatory gap between neurological theories and mysticism will ever be closed. Neurotheologians face not an explanatory gap but a chasm.

Sometimes, however, science can yield practical benefits in the absence of intellectual understanding. Quantum mechanics provides a case in point. Quantum theory raises more questions than it resolves about reality. Electrons and protons can act like waves or particles, depending on how we observe them, and their behavior appears to be both random and deterministic. But most physicists are oblivious of these conundrums. To them, all that matters is that quantum mechanics *works.* It predicts the outcome of experiments with astonishing accuracy. Directly or indirectly, quantum physics has yielded such powerful technologies as transistors, lasers, and nuclear reactors.

In the same way, perhaps the research that Austin calls perennial psychophysiology can yield practical applications in lieu of intellectual insights. This, in fact, is Austin's goal. He hopes that his book will inspire more research on the neurophysiology of spirituality, which in turn will lead to improved mystical technologies. The "real challenge" for neuroscience, Austin asserted in *Zen and the Brain,* should be to help more people reach "the full moon of enlightenment." (Here is another point of divergence between Austin and the neurotheologian Andrew Newberg, who expressed no interest in improving mystical technologies.) The widespread adoption of more effective spiritual practices might help society as a whole become less selfish and materialistic, more humane and creative. "That's my hope," Austin told me.

The transpersonal psychiatrist Arthur Deikman, whose work Austin discussed in *Zen and the Brain,* has coined the terms *instrumentality* and *automatization* to describe two basic cognitive tendencies that impede our mystical awareness. Instrumentality is our compulsion to view the world through the filter of our selfish interests. Automatization is our propensity to learn tasks so thoroughly that we perform them with little or no conscious thought. Some degree of instrumentality and automatization is necessary for survival. Automatization in particular is an attractive cognitive feature, because it allows you to carry out more than one task at the same time. You can ponder the mind-body problem while driving your kids to school, or fret over the slumping stock market while making love. But together, instrumentality and automatization conspire to transform us into virtual robots or zombies who lumber through life without ever really *seeing.*

The goal of Zen, Austin says, is to induce what Deikman calls *deautomatization,* a disruption of our routine ways of perceiving and interacting with the world. We do not see the world as a collection of objects to be manipulated; we simply *see* it. But as Austin acknowledged, for most people meditation, yoga, prayer, and all traditional spiritual practices usually work slowly, if they work at all. "Positive spiritual experiences do not always occur," he noted. "Indeed, for some meditators, the long-delayed 'awakening' is a profound 'bleak experience.'"

The Doors of Deception

In *Zen and the Brain,* Austin examined biofeedback, a once promising mystical technology that emerged from brain science in the 1960s. Biofeedback machines allow you to monitor your own electroencephalographic signals, or brain waves, and with practice to exert some control over them. After early studies suggested that deep meditative states have distinctive EEG signals, biofeedback was touted as a quick and easy way to achieve mystical mastery.

Unfortunately, later studies indicated that there are no electroencephalographic patterns unique to meditation or mysticism. As Austin noted, EEG signals vary from individual to individual. Moreover, the brain waves of an individual may shift back and forth between different rhythms during a single meditation session. Although psychotherapists and meditation teachers still prescribe biofeedback as a relaxation method, it has been largely discredited as a mystical technology.*

Zen and the Brain also reviewed research on a more popular mystical technology: psychedelic drugs, or entheogens, as Huston Smith calls them. Austin has never ingested these substances himself. He granted that psychedelics have the potential to induce mystical experiences — the Good Friday experiment and other studies seemed to have demonstrated as much — but he worried that their physiological and psychological effects are too multifarious and sometimes too dangerous to bring about healthy spiritual growth.

Austin feared that, unlike the slower-acting, endogenous excitotoxins released by meditation and other spiritual practices, which etch the brain in a precise, controlled fashion, psychedelic drugs may wreak havoc in the brain. Their effects on the brain's biochemistry seem to be "so complicated and interactive that the consequences are not only unknown but possibly unknowable," he said. Moreover, the hallucinations that accompany psychedelic experiences "are a distraction. You can be captured by them." In other words, psychedelics may be doors of deception, not perception.

In spite of these risks — and the legal prohibitions against psychedelics — a growing number of scientists, scholars, and religious leaders

---

*The science journalist Jim Robbins recently tried to rehabilitate the reputation of biofeedback, or a refined version called neurofeedback, in *A Symphony in the Brain* (Grove Press, New York, 2000). But the altered-states expert Charles Tart, who in the late 1960s thought biofeedback might prove to be an "electronic Zen," concluded in the most recent edition of his book *Altered States of Consciousness* (HarperCollins, New York, 1990, pp. 579–580): "I suspect that the more sophisticated use of biofeedback techniques will eventually aid basic meditative practices, but I doubt that it will substitute for it." However, Tart thinks virtual-reality devices may have promise as a mystical technology, for example, by simulating the *bardo* state that Tibetan Buddhists believe follows death. Virtual bardo, or "vardo," training could help us stay calm when we actually die, Tart said in an interview ("One Path or Many?" by Richard Smoley and Jay Kinney, *Gnosis*, Summer 1993, p. 35).

are quietly urging a reconsideration of their spiritual benefits. In the mid-1990s, some of these advocates of entheogens (the term they prefer) created the Council on Spiritual Practices to promote the cause of "making direct experience of the sacred available to more people" (as a brochure put it). The council's approach is geared toward education rather than lobbying. It maintains a Web site and sponsors the publication of books such as *Psychoactive Sacramentals,* a collection of essays edited by Thomas Roberts, the psychologist I met at the "Mystics" meeting in Chicago, and *Cleansing the Doors of Perception,* by Huston Smith.

As Smith pointed out in that book, humans have ingested mind-altering substances in spiritual contexts for millennia. What sets the modern era apart is that scientists now have the capacity to isolate and synthesize the active ingredients of psychotropic plants and fungi, to invent new psychotropic compounds, and to monitor their effects in the brain. The age of scientifically based psychedelic mysticism can arguably be traced back to 1898, when the German chemist Arthur Heffter showed that mescaline is the primary active ingredient of peyote; mescaline was synthesized some twenty years later. But most historians would probably agree that the modern psychedelic era really began in 1943, when a Swiss chemist discovered what remains, microgram for microgram, the most potent known mystical technology.

8

In the Birthplace of LSD

On April 16, 1943, Albert Hofmann was working at the pharmaceutical firm Sandoz, in Basel, Switzerland, when he was overcome by what he recalled later as "remarkable restlessness, combined with a slight dizziness." The thirty-seven-year-old chemist had been investigating a compound related to ergot, an extract of a fungus that infects grain-producing plants. Although ergot is poisonous, small doses have medicinal properties, such as inducing uterine contractions during childbirth and stanching postpartum bleeding. Hofmann hoped that this particular ergot compound, which he had originally synthesized five years earlier, might have potential for stimulating blood circulation.

Hofmann guessed that he had become dizzy after absorbing the compound through his skin. Three days later, to test his theory, he dissolved what he thought would be an extremely small dose — 250 millionths of a gram, or 250 micrograms — in a glass of water and drank it. Within forty minutes Hofmann felt so disoriented that he asked a colleague to escort him as he rode his bicycle home. He managed to make his way to his house, where "furniture assumed grotesque, threatening forms" and a kindly neighbor bringing him milk looked like a "malevolent, insidious witch with a colored mask." Hofmann feared he was losing his mind, or even dying. He was tormented by the thought that his wife and three children would never understand "that I had not experimented thoughtlessly, irresponsibly, but rather with the utmost caution."

Gradually, "the horror softened and gave way to a feeling of good fortune and gratitude." This sense of well-being persisted through the following morning. When he walked out into his garden after a rainfall, "everything glistened and sparkled in a fresh new light. The world was as if newly created." Thus, on April 19, 1943, now commemorated in some circles as Bicycle Day, Albert Hofmann discovered the psychotropic properties of lysergic acid diethylamide, his "problem child," as he called it later. Although its effects resembled those of mescaline, which had been synthesized more than two decades earlier, LSD was orders of magnitude more powerful. Just one hundred micrograms, or one ten-thousandth of a gram, produces intense effects.

Hofmann's contributions to psychedelic chemistry did not stop with LSD. In the late 1950s he showed that psilocybin — the compound that Huston Smith and other religious scholars ingested in the Good Friday experiment in 1962 — and psilocyn are the two primary active ingredients of *Psilocybe cubensis,* a mushroom consumed as a sacrament by Indians in Central and South America. At roughly the same time, the Hungarian-born chemist Stephen Szara, who later became a researcher at the U.S. National Institute on Drug Abuse, discovered the psychotropic effects of dimethyltryptamine (DMT), which had been synthesized in 1931, by injecting it into himself.

Scientists around the world were soon examining these compounds. LSD, mescaline, psilocybin, and DMT turned out to be remarkably similar in structure to chemicals found in the central nervous system. In fact, investigations of psychedelics helped reveal just how vital neurotransmitters such as dopamine, norepinephrine, and serotonin are to the brain's functioning. Researchers initially viewed LSD and related compounds as psychotomimetics, which mimic the symptoms of schizophrenia and other mental illnesses. The astonishing potency of LSD, in particular, implied that mental illness might be caused by subtle chemical changes in the brain. Eventually, psychiatrists began exploring the potential of LSD and similar drugs as treatments rather than simulants of mental disease. By the mid-1960s, peer-reviewed journals had published over one thousand papers describing the treatment of more than forty thousand patients with hallucinogens.

Drawing on this literature, *LSD: The Problem-Solving Psychedelic,* a book by two journalists published in 1967, touted LSD as a panacea that can "cure" schizophrenia, depression, autism, alcoholism, compulsive gambling, sexual frigidity, sexual "perversion" (the authors included homosexuality under this rubric), and other disorders. Citing the Good Friday experiment as well as personal testimonials, the book also contended that

LSD can induce profoundly transformative mystical experiences in healthy people.

At the same time, reports in the media and scientific literature linked LSD to suicidal, antisocial, and murderous behavior; long-term depression, anxiety, and psychosis; and genetic mutations that could cause birth defects. By the late 1960s state and federal officials alarmed about psychedelics' adverse effects — and their surging popularity among young people — passed laws prohibiting their use. Most legal investigations of psychedelics, and particularly those involving human subjects, soon ground to a halt.

In *Psychedelic Drugs Reconsidered,* first published in 1979 and updated in 1997, Lester Grinspoon and James Bakalar, professors of psychiatry and law, respectively, at Harvard, contended that both the benefits and the risks of psychedelics have been exaggerated. Studies claiming dramatic therapeutic results were for the most part flawed — lacking controls, with vague criteria for therapeutic success, and poor follow-ups — but reports on the dangers of psychedelics had even less basis in fact. Psychedelics occasionally cause persistent mental instability — including "flashbacks," triplike symptoms recurring long after the drug's immediate effects have subsided — but these cases are rare, Grinspoon and Bakalar asserted.* They concluded that the pros and cons of psychedelics can be evaluated only with a new generation of rigorous, objective studies.

Since the early 1990s, there has been a quiet resurgence of such research. One instigating force is the Heffter Research Institute, founded in 1993 by several scientists with an interest in psychedelics. Named after the German chemist who discovered mescaline in 1898, the institute is based in

---

*Grinspoon and Bakalar defined "flashback" as "the transitory recurrence of emotions and perceptions originally experienced while under the influence of a psychedelic drug . . . They usually decrease quickly in number and intensity with time, and rarely occur more than a few months after the original trip" (*Psychedelic Drugs Reconsidered,* Lindesmith Center, New York, 1997, p. 159). Studies carried out in 1972 found that roughly one in four users of psychedelics had flashbacks; one survey found that 57 percent of those who had flashbacks considered them to be pleasant, but 32 percent viewed them as "somewhat frightening" and 11 percent as "very frightening" (pp. 160–161). People who suffered from severe, prolonged reactions to psychedelics usually differed from schizophrenics in several ways: The drug users tended to have visual rather than auditory hallucinations, they did not display the dulled emotions typical of chronic schizophrenia, and they had more insight into their condition (p. 164). Curiously, Prozac and similar antidepressants, which like LSD affect levels of serotonin in the brain, have reportedly caused flashbacks in former psychedelic users.

Santa Fe, New Mexico. According to James Thornton, executive director of the institute, it seeks to "find new ways of treating illnesses such as depression and addiction; find ways to improve well-being, including the quality of life for the terminally ill; and understand the basic science of how the brain, mind, and consciousness work." The Multi-disciplinary Association for Psychedelic Studies (MAPS), founded in 1986 and based in Sarasota, Florida, supports research on the classic psychedelics, such as LSD and psilocybin, as well as other psychotropic drugs, such as marijuana. According to the masthead of the *MAPS Bulletin*, a quarterly journal, the organization's ultimate goal is "gaining government approval for [psychedelic drugs'] medical uses."

The founder of MAPS, Rick Doblin, became an advocate for drug-law reform as a result of his own youthful psychedelic experiences. He has emphasized the need for renewed research by publishing critiques of two legendary pro-psychedelic studies. One was the Good Friday experiment (which was described, along with Doblin's analysis, in chapter one). Doblin has also examined the so-called Concord Prison experiment. Carried out in the early 1960s by Timothy Leary and several Harvard colleagues, the experiment involved thirty-two inmates of the Massachusetts Correctional Institute in Concord who were approaching their release dates. Over a six-week period, each inmate received psilocybin twice and participated in bi-weekly group psychotherapy sessions. Leary and his colleagues claimed that after the inmates were released, their recidivism rate was much lower than that of convicts with similar backgrounds. "We had kept twice as many convicts out on the street as the expected number," Leary wrote in 1968. After examining Leary's notes and prison records, Doblin concluded that the recidivism rate of Leary's subjects was actually slightly higher than that of other convicts.

Doblin said his critiques were intended not to discredit psychedelic research but to highlight the need for careful, objective evaluations. "Leary's misleading reports serve as an object lesson in what not to repeat," he stated. "With the current renewal of research into the therapeutic use of drugs after three decades of almost total prohibition, psychedelic researchers must hold themselves to the highest ethical standards in order to retain a measure of trust from regulators and the general public."

"Worlds of Consciousness"

To gauge the current state of psychedelic research, in November 1999 I flew to Basel, the birthplace of LSD, to attend "Worlds of Consciousness," the Third International Congress of the European College for the Study of Consciousness. The European College's Web site calls it "a multi-disciplinary forum of scientists and scholars from many European countries. We have in common an interest in research into consciousness in its many possible states and conditions." I attended the congress because it had been described to me as a leading symposium on mystical technologies, particularly the psychedelic variety.

The conference assembled in an ultramodern convention center in downtown Basel. In a lobby just outside the main auditorium, vendors peddled psychedelic books, music, art — and plants. One vendor's counter displayed a potted peyote cactus festooned with wrinkled buttons; a terrarium sprouting dozens of beige-capped mushrooms, which a label identified as *Psilocybe cubensis;* a bowl containing chunks of a vine called *Banisteriopsis caapi,* one of ayahuasca's ingredients; a bundle of black leaves from the potent psychotropic plant *Salvia divinorum;* and the root of the African shrub *Tabernanthe iboga,* source of the hallucinogenic alkaloid ibogaine. Are these for sale? I asked a lank blond youth standing behind the counter. Yes, he replied with a German accent, and an implied "Duh."

The hall also featured an exhibit of psychedelic drawings by the Swiss surrealist H. R. Giger, best known for designing the creepy, organo-robotic sets and monsters of the *Alien* movies. Giger's drawings here showed pouty-lipped, warhead-breasted, cybernetic vixens flayed, tortured, transmogrified by unimaginable psychic forces. Brain cases mutated into rib cages — or were they radiators? — and fingers into voluptuous legs. Although some faces expressed agony, most were ecstatic or — more unsettling — serene. Is this how Saint Teresa of Ávila felt during her mystical betrothal to God, when He graced her with what she called the "gentle yet penetrating touches of His love"?

In contrast to this subjective evocation of the psychedelic experience, the scientists' lectures — which described their attempts to quantify the effects of LSD and psilocybin with positron emission tomography, electroencephalography, blood tests, and psychological questionnaires — seemed almost parodically dry. Although many of these scientists confessed privately that they had firsthand knowledge of psychedelics' transformative power, for the most part they concealed their fervor beneath clinical rhetoric.

One prominent speaker was David Nichols, chairman of the Department of Medicinal Chemistry at Purdue University and a cofounder of the Heffter Research Institute. He is an imposing figure: tall, stocky, with a trim white beard and sober demeanor. Because he works with animals rather than human subjects, Nichols is one of the few scientists in the world who have studied psychedelics continuously since the 1970s. He has trained rats to signal when they recognize the effects of such mind-altering drugs as LSD, methamphetamine, and MDMA. In this way, Nichols can determine whether novel compounds cause similar subjective sensations in rats, and hence potentially in humans. He can also determine the drugs' physiological effects by performing experiments on animals that would not be possible on humans.

One long-term focus of Nichols's current research is MDMA. Although chemically similar to mescaline (both are members of a class of compounds called phenethylamines), MDMA produces effects subtly different from those of the classic psychedelics, as Nichols and the chemist Alexander Shulgin first pointed out in a scientific paper in 1978. In 1986 Nichols coined the term *entactogen,* from the Greek and Latin roots for "touching within," to refer to MDMA's ability to reduce anxiety and enhance self-awareness and empathy for others.

Nichols believes that entactogens have great potential as adjuncts to psychotherapy, but he is concerned about their toxicity; his animal studies have turned up evidence that repeated doses of MDMA can damage serotonin receptors. Nichols hopes that continued research can identify new entactogens and psychedelics with fewer side effects. He was cautiously optimistic about an analog of LSD that supposedly produces its psychotropic effects but with a greatly reduced risk of panic. Some enthusiasts — but not Nichols — were calling it "super-LSD."

One up-and-coming researcher in Basel was John Halpern, a thirtyish, fast-talking psychiatrist at Harvard Medical School. Together with Harrison Pope, also a Harvard psychiatrist, Halpern has scoured the psychedelic literature for reports on adverse effects. They tentatively concluded in a 1999 paper that "there are few, if any, long-term neuropsychological deficits attributable to hallucinogen use." The panic and psychosis that characterize bad trips usually fade after the drug wears off; users who suffer long-term psychosis or other disorders generally had a preexisting mental illness or took other, more harmful drugs in addition to psychedelics, Halpern said.

To better assess the effects of protracted consumption of psychedelics, Halpern and Pope have undertaken a study of members of the Native American Church, who consume peyote as a sacrament. According to preliminary results that Halpern presented in Basel, church members show no adverse psychological or physiological effects from peyote. In fact, overall they are healthier and happier — and much less prone to alcoholism — than non–church members. Halpern was careful to point out that these benefits could derive from the social fellowship provided by church membership. He nonetheless argued that psychedelics, and especially short-acting ones such as DMT, should be reconsidered as treatments for substance abuse and other disorders.

Evgeny Krupitsky, a psychiatrist who heads a substance-abuse clinic in St. Petersburg, Russia, reported that for almost two decades he has successfully treated alcoholics with ketamine supplemented by individual and group psychotherapy. Ketamine is a general anesthetic — used more often in veterinary than human medicine — that when injected at subanesthetic doses triggers a disorienting hallucinogenic episode lasting an hour or so. In one of Krupitsky's studies, 73 out of 111 alcoholics, or almost 66 percent, abstained from drinking for at least a year after their ketamine treatment, compared to 24 percent of those in a control group.

The ketamine experience can be ego-shattering, Krupitsky said, but that, in a sense, is the point. Ideally, patients feel such revulsion for their alcoholic selves that they strive to re-create themselves as nonalcoholics. One trick the therapists employ is to make the ketamine-intoxicated patients sniff a bottle of booze at the peak of their session; the patients' disgust often persists long after the ketamine's effects have worn off. Like John Halpern, Krupitsky was careful to qualify his results. He noted that some alcoholics decline ketamine therapy after being forewarned about its harrowing aspects; those who undergo the treatment may be more highly motivated to stop drinking than alcoholics in the control group.

An entire afternoon session in Basel was devoted to ayahuasca, which was apparently becoming the entheogen of choice for spiritual seekers with a taste for the exotic. This mysterious Amazonian potion — whose other

Indian names are yage, caapi, natema, pinde, and karampi — is made by boiling two plants together for as long as a week. It has long played a role in the religious and cultural lives of Amazonian Indians and mestizos, who are of mixed racial ancestry.* Ayahuasca now serves as a sacrament not only for indigenous Indians but also for several syncretic religious movements, which combine elements of Catholicism, esoteric European traditions, and Afro-Brazilian and Amazonian spirituality. The two largest ayahuasca churches, both based in Brazil, are União do Vegetal, which claims six thousand members, and Santo Daime, which reportedly has five thousand members. Underground offshoots of these churches are springing up in the United States and Europe.

The Beat writer and drug omnivore William Burroughs vividly described ayahuasca's effects in correspondence with Allen Ginsberg that was published in 1963. Burroughs said that minutes after a shaman gave him ayahuasca in an Amazonian village in 1953, he became violently ill. He spent the evening in a fearful fever dream, convinced he had been poisoned. He imagined himself transformed into a black woman, then an androgynous man-woman. "All I want is out of here," he said over and over. Later, Burroughs had more pleasant ayahuasca trips, and he praised the drug's ability to facilitate "space time travel."

Burroughs had learned about ayahuasca from the renowned Harvard botanist Richard Shultes, who collected samples of ayahuasca's constituent plants and analyzed them in the 1940s. Shultes, who first described aya-

---

*For a fascinating overview of ayahuasca shamanism, see *Ayahuasca Visions* (North Atlantic Books, Berkeley, California, 1991). It consists of a scholarly essay by Luis Eduardo Luna, a Colombia-born anthropologist and authority on ayahuasca who leads seminars on ayahuasca in Brazil, followed by forty-nine paintings by the Peruvian shaman turned artist Pablo Amaringo. In captions for the paintings, which show jungles teeming with jaguars, snakes, demons, angels, and aliens in spaceships, Amaringo recalls the ayahuasca visions that inspired them. The paintings have received radically different treatments in two recent books. In *Consilience* (Knopf, New York, 1998), Edward O. Wilson contended that one day all aspects of human culture — including art and religion — will yield to conventional scientific explanation. As an example, Wilson argued that the snakes recurring in Amaringo's fantastical paintings — and in a great deal of indigenous art — have a genetic basis; they are products of our innate herpetophobia, which natural selection embedded in our brains many millennia ago. In *The Cosmic Serpent* (Jeremy Tarcher/Putnam, New York, 1998), the anthropologist Jeremy Narby conjectured, after encountering snakes in his own ayahuasca session, that they are actually embodiments of DNA, the double helix, which is a "minded," or intelligent, life form that came to earth millions of years ago and seeks to communicate with us in our dreams and trances.

huasca in print in 1951 and consumed the drug himself more than twenty times, highlighted the great mystery posed by ayahuasca: How did the Indians figure out that two plants that are marginally psychoactive when consumed individually become intensely psychoactive when consumed together? By all accounts a hard-nosed rationalist, Shultes said he could not discount the Indians' claim that their ancestors learned how to brew the sacred potion from the spirits of the ayahuasca plants.

Although different plants are occasionally used, the two species most commonly associated with ayahuasca are a bush called *Psychotria viridis* and the vine *Banisteriopsis caapi*. *Psychotria viridis* contains dimethyltryptamine, or DMT. When pure synthetic DMT is injected or smoked, its effects can last from ten minutes to half an hour. Ingested orally, DMT is usually rendered inactive by an enzyme in the gut called monoamine oxidase. But *Banisteriopsis caapi* contains the alkaloids harmine and harmaline. These are monoamine oxidase inhibitors, which block the effects of monoamine oxidase. What all this means is that when *Banisteriopsis caapi* and *Psychotria viridis* are boiled together, the resulting brew delivers DMT in an orally active form. The trouble is, the stomach enzyme monoamine oxidase, which renders DMT inactive, also protects us from toxic reactions to proteins found in many foods.

In spite of ayahuasca's unpleasant, sometimes dangerous side effects (which are spelled out in chapter eleven), at least one study has shown long-term consumption to be relatively safe. An examination of members of the União do Vegetal (UDV) carried out in the early 1990s found no adverse physiological effects from prolonged use of ayahuasca. Church members also were less prone to alcoholism, drug abuse, and violent behavior than they had been before they joined the church. Tests suggested that these benefits might have a physiological basis; UDV members had elevated levels of serotonin, which have been correlated with greater self-esteem and reduced aggression.

One enthusiastic proponent of ayahuasca is the German anthropologist Christian Ratsch, who was one of the organizers of the Basel meeting. If tie-wearing scientists such as David Nichols, John Halpern, and Evgeny Krupitsky represented the responsible, rational superego of the psychedelic-research community in Basel, Ratsch was its id. Although most of his books and articles are in German, he is known among the cognoscenti as an expert on the varieties of psychedelic shamanism. Ratsch has reportedly never cut his raven-black hair and beard, and in Basel he was clad entirely in black leather: pants, hat, boots, fringed jacket. His slouch and half-lidded eyes gave him a reptilian air, which his gravelly, thickly accented croak did nothing to dispel.

Ayahuasca, Ratsch warned me, is "definitely *not* a recreational drug," but it is also "definitely the best shamanic medicine ever." Indian shamans, he said, view ayahuasca trances primarily as a means to an end, not an end in themselves. They believe that under the influence of ayahuasca they acquire magical powers with which they can heal their friends, harm their enemies, and learn about the properties of other plants. But Ratsch offered a different reason for using the stuff: "You take a good dose of ayahuasca, and then you're enlightened."

Ratsch defined enlightenment as "a state of complete understanding of everything, that total loss of ego structures for a while, and just being one with everything." He denied the contention of mystical experts such as Ken Wilber that decades of meditation or other spiritual practices are required to achieve enlightenment. "That's such a *bad lie,* and an exploitation of needs," Ratsch growled. Enlightenment "has nothing to do with all these spiritual teachings"; it merely requires "the right molecule to hit your brain." Enlightenment is a transitory state, like an orgasm. In fact, Ratsch said, some Amazonian societies use their term for orgasm to describe drug-induced raptures. "You are not in a permanent state of orgasm. It's one peak and then you have to recharge your batteries." The same is true of enlightenment.

Enlightenment does not give you any kind of final knowledge, Ratsch said. Quite the contrary. You see that the search for truth and meaning is endless. "If the search for knowledge stops, then you're basically" — Ratsch paused — "dead as a living, exploring being." The universe "produces people like us to learn about itself." This self-exploring process "goes on and on and on. And nobody knows where it goes and what happens. And I think that's part of enlightenment, to understand that there *is* no aim."

The cautious psychedelicist

I heard a more sober assessment of the psychedelic experience from the Swiss psychiatrist Franz Vollenweider. Based at the University of Zurich Medical School and at the Psychiatric University Hospital, where Carl Jung once worked, Vollenweider is arguably the world's leader in psychedelic research involving humans. He has been remarkably successful in obtaining financial and political support for his studies and in getting his results published in peer-reviewed journals. One reason may be that he is so resolutely unpsychedelic.

Slender, neither tall nor short, with a trim mustache and longish hair graying above the ears, Vollenweider is unremarkable in appearance. He is mild-mannered, soft-spoken, prone to dry scientific abstractions. His research, he told me, has three major goals: understanding the neurobiological and cognitive effects of psychedelics, understanding the neurobiology of mental illnesses such as schizophrenia and finding better treatments for them, and studying the efficacy of psychedelics as aids to psychotherapy. One of his most ambitious projects involves using brain-imaging machines to map the neural circuitry activated by different drugs. Eventually, these studies might reveal how the physiological effects of psychedelics resemble and differ from the effects of meditation, schizophrenia, and other mind-altering influences.

Vollenweider has tentatively confirmed findings first reported by the German psychologist Adolf Dittrich in the 1980s (which I mentioned in the introduction). Dittrich interviewed more than fifteen hundred healthy subjects in six countries whose consciousness had been altered by drugs, meditation, hypnosis, sensory deprivation, sensory stimulation (exposure to intense images and sounds), and other means. He also examined people whose consciousness was impaired by schizophrenia and other illnesses. No matter how the altered states are produced, Dittrich concluded, they all fall into three broad categories, or "dimensions": oceanic boundlessness, dread of ego dissolution, and visionary restructuralization, otherwise known as heaven, hell, and visions.

Vollenweider observed this same pattern in the experiences of his subjects. By scanning their brains, he found that each dimension is associated with a different pattern of neural activity. To simplify a bit: Heaven is correlated with increased metabolic activity in the frontal and parietal lobes, regions at the front and top of the cortex, respectively, that underpin reasoning, language, and other higher cognitive functions. Hell corresponds with heightened activity in the thalamus, a walnut-sized structure deep in the brain that regulates emotion. Visions correlate with increased activity in the striatum, a cluster of nerves near the base of the brain that processes raw sensory data, and with suppressed activity in the occipital lobes, an area at the rear of the brain that underpins visual perception.

Vollenweider's findings do not mesh neatly with those of other investigators of mysticism. Whereas he links the classic, "heavenly" mystical experience to increased activity in the parietal lobes, for example, Andrew Newberg has speculated that activity in these regions decreases during mystical states. And unlike Michael Persinger, Vollenweider has found little evidence that the temporal lobes play a crucial role in mediating altered states. Of course, these divergences may reflect the fact that Vollenweider focuses on states induced by drugs rather than by meditation or electromagnetic pulses.

He emphasizes, moreover, that all findings about the neurophysiology of mystical states should be interpreted cautiously, given the complexity of these states and the crudeness of brain-imaging technology. Even the sophisticated PET machine at his hospital in Zurich takes a minute to scan a brain; it thus cannot track the rapid changes in subjective experiences that often characterize a psychedelic session. New magnetic-resonance-imaging machines have a much better time resolution, but they are so noisy that they disrupt subjects' thoughts. A technique called high-resolution electroencephalography can detect millisecond-long fluctuations in brain waves, but it yields no information about activity deep in the brain. "Every method has its pros and cons," Vollenweider said.

He was just as candid in discussing the pros and cons of psychedelics. He and other researchers have found that low doses of psilocybin can enhance learning and memory. Brain scans and psychological tests also suggest that psilocybin helps subjects tap into a larger "semantic network" when making associations between words and concepts; this phenomenon may account for anecdotal reports from artists, musicians, and writers that psychedelics spur creativity. Vollenweider believed that psychedelics and entactogens can also serve as psychotherapeutic catalysts, helping patients to understand themselves and uncover their capacity for positive emotion. He hoped someday to conduct clinical trials of psilocybin, MDMA, and other drugs for the treatment of anorexia nervosa, obsessive-compulsive disorder, and depression.

But Vollenweider warned that all these drugs pose risks. The prospects for psychedelic therapy would be improved if chemists could find drugs with fewer side effects, such as entactogens that resemble MDMA but are less toxic. He was dismayed that some of his more overtly pro-drug colleagues downplay the adverse effects of mind-expanding compounds. At a recent conference, one advocate had pressured Vollenweider to proclaim the safety of MDMA. Vollenweider refused to do so, because his own research and anecdotal reports from users led him to suspect that long-term use of MDMA can indeed damage neural receptors. (He has seen no evidence, however, to support claims that a single dose of MDMA can cause irreversible harm.)

Before giving a drug to a subject, Vollenweider makes a point of taking it himself in a research setting. He thus has direct knowledge of the negative psychological consequences of psychedelics. During one psilocybin session, he briefly feared he was going mad or dying, and he was overcome by horror at the thought of losing those he loved. These nightmare visions "appear so real, superreal. If you are not stable, I could imagine that some people in uncontrolled settings could go crazy," he said.

Vollenweider acknowledged that at this point the research is raising more questions than it answers. One puzzle concerns the effect of psychedelics on the startle response, the body's physiological reaction to an abrupt, intense stimulus such as a loud noise. Researchers showed long ago that when a weak stimulus immediately precedes a strong one, the body has a decreased startle response to the strong stimulus. This phenomenon is called pre-pulse inhibition.

Schizophrenics typically display weaker-than-normal pre-pulse inhibition. This trait is thought to contribute to the sensory overload characteristic of schizophrenia. Experiments have shown that psychedelics disrupt pre-pulse inhibition in animals, but the evidence from humans is confusing. In some studies psychedelics increase pre-pulse inhibition, and in others there is a decrease. "We cannot explain what it all means at the moment," Vollenweider said.*

*Research on the relation between meditation and habituation (which is closely related to pre-pulse inhibition) has produced similarly ambiguous results, as both James Austin and Susan Blackmore have pointed out. Whereas some veteran meditators are acutely responsive to external stimuli, others are utterly *un*responsive. Mystical technologies seem in some cases to sensitize us to the world and in other cases to insulate us from it.

He nonetheless hopes that research on psychedelics will lead to better understanding of and treatments for mental illness. Some of his funding comes from pharmaceutical companies who share this hope. In fact, Vollenweider has resurrected the old notion that psychedelics are psychotomimetics, which mimic psychosis. But will psychedelics ever be accepted as sacraments or catalysts for spiritual growth, I asked, if they are so closely associated with mental illness?

To my surprise, Vollenweider seemed to doubt whether psychedelics *should* be accepted as sacramental substances. Although drugs can precipitate deeply meaningful experiences, "if you come up and label that sacred or religious, for some people it could just be an artificial construction, not a real breakthrough." He worried that religious organizations that encourage the use of drugs and provide ready-made interpretations of the experiences could turn into cults.

Vollenweider looks skeptically at drug-induced visions, including his own. He had visions in which he turned into animals, including a fish swimming in the sea, and once felt as though he were reliving the evolution of life. But he was inclined to interpret these events as creations of his own psyche rather than as revelations from some cosmic, transpersonal realm. During one trip, he had a heavenly experience in which he felt love and bliss suffusing himself and the entire universe. Rather than embracing this as absolute truth, he looked at it with the cool eye of a scientist.

"It was real in my experience, but it was self-created, and it also satisfies my wishes," Vollenweider said. "I cannot expand it to an intelligence that just wants everybody so happy." His blissful feeling was similar to the one that Huston Smith had during the Good Friday experiment. But unlike the American religious scholar, the Swiss psychiatrist chalked up his vision of a loving God to anthropomorphism and wishful thinking.

Albert Hofmann's second thoughts

Albert Hofmann, who inaugurated the modern psychedelic era by discovering LSD's effects in 1943, has for the most part been a vigorous protector of his legacy. Although he turned ninety-three in 1999, the year of the "Worlds of Consciousness" conference in Basel, Hofmann was still avidly following the field he helped create. A stooped, white-haired man in coat and tie with a fierce Churchillian mien, he attended virtually all the scientific sessions at the meeting. He was also the guest of honor at an evening gala emceed by the ayahuasca evangelist Christian Ratsch.

When I first approached him, he was reluctant to speak to me. In halting, heavily accented English, he explained that after more than half a century he was weary of answering questions about LSD. But one day we spoke during the lunch break, and Hofmann energetically defended his "problem child." Timothy Leary, whom Hofmann had met several times, was partly to blame for LSD's acquiring such a bad reputation. "I had this discussion with him. I said, 'Oh, you should not tell everybody, even the children, "Take LSD! Take LSD!"'" LSD is too powerful for young people, including teenagers, Hofmann said. "They are still in growth, and it is a very dangerous stage."

LSD is "very, very potent, and everything that is potent is dangerous." If used improperly, LSD "can hurt you, it can disturb you, it can make you crazy." But Hofmann was outraged that scientists and psychiatrists are still prohibited from investigating and prescribing it in a safe, controlled fashion. "That is nonsense! Absolute nonsense! I don't want to promote absolute freedom," Hofmann said. "But the medical professions should have access to it."

Although it can harm people by provoking reckless or suicidal behavior, LSD is neither toxic nor addictive, he said, and it has never killed anyone by overdose. Used with respect, it has enormous potential as a tool for investigating human consciousness and in psychotherapy. Psychedelics can also stimulate the "inborn faculty of visionary experience" that we all possess as children but lose as we mature. Hofmann hoped that in the future people would be able to take psychedelic drugs in "meditation centers" to awaken their religious awe.

Hofmann saw no incompatibility between the scientific and spiritual outlooks. "They are not contradictory. They are complementary." Reality, he said, has both material and spiritual aspects; science addresses the material aspect and religion the spiritual aspect. Conflict occurs when the scientist insists that the world is entirely material, or, conversely, when a religious person insists that the material realm is an illusion and only spirit matters. At their best, both religion and science can help us appreciate "how wonderful creation is," Hofmann said. "We rarely see it! We forget it!"

I asked Hofmann if he believes in God, or some form of intelligence or plan underlying reality. "I am absolutely convinced," he replied, "by feeling and by knowledge — my knowledge as a natural scientist — that there must be a creative spirit, an intelligence, which is the reason for what we have." Everything that exists, Hofmann said, pounding the table between us with his fist, is a manifestation of this plan. "It is impossible to have this without a plan. Otherwise you have only material, material, material!"

Hofmann has had frightening psychedelic experiences, including the early stages of his first trip, in 1943, but they invariably yielded to positive emotions. His worst trip occurred under the influence of psilocybin, when he hallucinated that he was in a ghost town deep in the earth. "Nobody was there. I had the feeling of absolute loneliness, absolute loneliness. A terrible feeling." When he emerged from this nightmare and found himself with his companions again, he felt ecstatic. "I had the feeling of being reborn! To see now again, and see what a wonderful life we have here!" The gruff old man looked past me, his eyes gleaming, as if reliving that moment.

Hofmann has asserted that hellish trips have as much to teach us as heavenly ones. "Only when we are conversant with both heaven and hell is our life full and rich," he once wrote. "And it is fuller and richer the more deeply we experience both." But Hofmann also confessed misgivings about psychedelics' destructive power. Some of the young LSD enthusiasts who had appeared at his doorstep over the years were obviously troubled.

Writing to a friend in 1961, Hofmann divulged misgivings about having brought LSD and psilocybin into the world. He feared that psychedelics might "represent a forbidden transgression of limits." He compared his discoveries to the discovery of nuclear fission; just as nuclear fission threatens our fundamental physical integrity, so do psychedelics "attack the spiritual center of the personality, the self."

Hofmann spelled out what he felt were the most disturbing metaphysical implications of LSD. The fact that minute amounts of a chemical can have such profound effects on our perceptions, thoughts, and beliefs suggests that free will, which supposedly gives us the power to shape our destiny, might be an illusion. Moreover, our deepest spiritual convictions may be nothing more than fluctuations in brain chemistry. To emphasize this point, Hofmann quoted from an essay he had recently read: "God is a substance, a drug!"

Synesthesia, Form Constants, and Ultramaterialism

This is the paradox of psychedelic research, that it is quite compatible with a crushing materialism. Scientists have already provided straightforward, reductive explanations of some of psychedelics' extraordinary subjective effects. For example, they often cause synesthesia, which has been estimated to occur naturally in only one out of every two thousand people. This condition causes input from one sense to be perceived as if channeled through another sense. You perceive a guitar riff as a flock of brightly colored birds, say, or sunlight filtering through leaves as the strumming of a harp. Experiments on rats have shown that LSD and other psychedelics stimulate neural activity in the locus coeruleus, a cluster of nerve cells in the brain stem through which incoming signals from all five senses are funneled before being routed to separate processing centers. These experiments suggest that synesthesia — at least in its drug-induced form — might result from short-circuiting between the different sensory channels in this neural bottleneck.

The explanation for another peculiar perceptual effect can be traced back to observations first made in the 1920s by Heinrich Kluver, a neuroscientist at the University of Chicago. After taking mescaline himself and interviewing others who had taken the drug, he concluded that its visual distortions comprise four geometric patterns: lattices, spirals, tunnels, and cobwebs. Kluver noted that these "form constants" are triggered by many conditions besides drug intoxication, including migraine headaches, epileptic seizures, psychosis, dreams, delirium tremens, high fever, sensory deprivation, and oxygen deprivation.

Recently another University of Chicago researcher, the mathematical biologist Jack Cowan, proposed that form constants stem from basic neurophysiological processes. Cowan's theory depicts the brain as a kind of cellular automaton. In its simplest form, a cellular automaton is a grid of squares, or cells, that can be one of two colors. A set of rules relates the color of each cell to the color of its immediate neighbors. A change in the color of one cell triggers a cascade of changes throughout the system. Even cellular automatons based on very simple rules can generate fantastically complex effects.

Psychedelics, Cowan notes, are known to boost the random discharge of neurons in the visual cortex, just as epileptic seizures and other conditions associated with form constants do. As each neuron discharges, it makes adjacent neurons more likely to discharge as well. As waves of neural excitation propagate through the visual cortex, they produce effects similar to those generated by cellular automatons, notably the roiling, convective dynamism and form constants characteristic of psychedelic visions. Cowan has demonstrated his theory by constructing detailed computer simulations of the visual cortex. His simulations can reproduce not only form constants but also the multiple images, expanding images (megalopsy), and shrinking images (micropsy) often reported by users of psychedelics.*

Ronald Siegel, a UCLA psychopharmacologist, showed in the 1970s that even pigeons perceive form constants when administered psychotropic drugs. Siegel trained pigeons to distinguish between form constants in an unintoxicated state; the pigeons could then signal which form constant they were perceiving while under the influence of a drug. After interviewing hundreds of drug users, Siegel concluded that certain themes also recur in the more complex, dreamlike hallucinations induced by psychedelics. Almost half of the subjects perceived small animals or humanlike creatures, "many in the nature of cartoons and caricatures," Siegel wrote in *Scientific American* in 1977.

Moreover, almost three quarters of Siegel's subjects reported seeing images with religious significance, such as crosses or mandalas. Siegel predicted that these more complex hallucinations will someday be explained in neurophysiological terms, just as form constants and synesthesia have been. Indeed, researchers such as Franz Vollenweider — whose PET scans are revealing the neural basis of heaven, hell, and visions — are already advancing toward that goal.

But this "ultramaterialism" has been vehemently rejected by one pioneer of modern psychedelic research, a psychiatrist whose work has been

---

*Geometric visual hallucinations may also result from direct electrical stimulation of the brain, as the Canadian neurosurgeon Wilder Penfield showed in the 1950s, and from radiation. Flashes of light reported by astronauts in early space missions are believed to have been triggered by cosmic rays striking the visual cortex; the earth's atmosphere normally shields us from such rays. A friend (now deceased) told me that while receiving radiation treatment for a brain tumor affecting his vision, he experienced visual hallucinations that reminded him of youthful psychedelic trips. At one point, he saw a vast stained-glass window shattering into infinite shards. He said the hallucinations were quite pleasant, even thrilling, in spite of their negative context.

cited by mystical experts as disparate as Huston Smith, Ken Wilber, James Austin, and the Catholic theologian Bernard McGinn. Although he began his career as a strict Freudian and an atheist, over the years his views changed radically as a result of his research on LSD and other drugs. Eventually, he left psychoanalysis and the materialistic paradigm of science far behind.

9
God's Psychoanalyst

Navigating the capillary-like byways of Marin County, a haven of spiritual seekers and teachers just north of San Francisco, I was slightly nervous. I had an appointment with one of the county's legendary residents, the psychiatrist Stanislav Grof, and I was less familiar with his views than I would have liked. I knew the basic facts of Grof's career. He was born and educated in Czechoslovakia and first investigated LSD there in the 1950s. He emigrated to the United States in 1967, just before the Soviet invasion of his homeland, to take a position at Johns Hopkins University, a prestigious center of psychiatric research.

There Grof tested LSD's potential for treating disorders such as alcoholism, drug addiction, and depression, and he administered the drug to terminally ill cancer patients to help them come to terms with their impending deaths. He observed LSD's effects on artists, writers, and mental health professionals, including himself. In 1973 he left mainstream academia to teach at Esalen, the cradle of the human potential movement. Later he became an instructor at the California Institute for Integral Studies, a center for holistic or alternative — some would say New Age — education based in Palo Alto.

Grof is a seminal figure in transpersonal psychology, the same field with which Ken Wilber is associated. In the late 1960s, Grof banded together with like-minded colleagues to create a psychology expansive enough to embrace the dimensions of human experience revealed by LSD and other mystical technologies. At Grof's suggestion, they called their psychology *transpersonal,* a term that Jung had coined to describe experiences in which we transcend our individual identities and tap into a "collective unconscious" shared by all humans. (Grof has called Jung "the forefather of transpersonal psychology.")

A tenet of the field is that transpersonal experiences have therapeutic potential. For obvious legal reasons, most transpersonal psychotherapists now advocate nonchemical methods for achieving transpersonal states. In the 1970s Grof and his second wife, Christina, developed a technique called holotropic breathwork, in which patients hyperventilate while lying down and listening to music. The Grofs, who oversee an international network of holotropic breathwork therapists, claim the technique can trigger altered states similar to those induced by drugs.

Grof had been difficult for me to track down. He did not return e-mail messages and phone calls I left over a period of months at the office that administers his holotropic technique. Finally, just before I traveled to San Francisco on other business, he called to say I could visit him at his home. As a result, I had time to read only one of his many books, *LSD Psychotherapy*, a relatively cautious, clinical discussion of LSD's therapeutic potential, published in 1980. The book revealed Grof's fascination with the birth-related images that he said often arise in LSD and holotropic sessions. The trauma of birth plays a key role in shaping the human psyche, Grof proposed, and reliving this trauma can be cathartic. Perinatal experiences also serve as a portal connecting the personal and transpersonal realms.

A few items in the book gave me pause. Grof said he witnessed "dramatic improvement" in an obsessive-compulsive man after an LSD session in which he lost control of his bowels and played with his feces for hours. Grof mentioned a 1958 paper by researchers who gave schizophrenics LSD followed by electroshock therapy; he seemed to accept the paper's claim that these poor souls improved. This and other studies cited by Grof recalled the Central Intelligence Agency's infamous Bluebird and MK-ULTRA programs, which in the 1950s and 1960s investigated the use of LSD as a brainwashing tool. Grof certainly acknowledged LSD's "negative potential"; he likened the drug to a knife, which can kill or heal depending on how it is used. He mentioned but — as far as I could tell from my quick read — did not pass judgment on the conviction of many subjects that under LSD's influence they acquired paranormal powers such as telepathy.

Leaving Freud behind

Arriving at Grof's driveway twenty minutes early, I pulled my car to the side of the road to go over my questions. Minutes later, a family drama unfolded before me. A man in tennis whites charged down the street toward a

Mercedes SUV parked across from me, shouting over his shoulder, "Hurry up!" Hustling behind him were a scowling woman and two scowling boys in baseball uniforms. As the woman approached the car, the man stuck his hand out and bellowed, "Give me the keys!" The woman frantically patted her shorts, her expression shifting from rage to horror to rage again. "I lost them running after you!" she shrieked, at which point man, woman, and boys all yelled at one another at once.

So furious was this fracas that I had an irrational fear they would turn on me and scream, "It's your fault!" I started my car and crept into Grof's driveway. I was still slightly rattled as I pulled up to his house, an ultramodern, geometric structure with pyramidal skylights, nestled in a cloud of trees. Greeting me at the door, Grof was stolid, sleekly handsome, his glossy dark hair combed straight back. He spoke in a preternaturally calm, deep, sonorous voice that bore traces of his Czech upbringing. He led me up a stairway past a vaguely vaginal painting to an office lined with books, including the multivolume *Encyclopedia of Religion* and the collected works of Aristotle, Freud, Jung, and Ken Wilber. Primitive stone figurines filled a glass case. On the desk stood a statue of the Hindu elephant god Ganesh, a photograph of Grof with Albert Hofmann, and a dark stone obelisk. Some sort of prehistoric sexual aid? I didn't ask.

Pressed up against the office windows was an enormous tree, its branches as sinuous and tangled as dendrites. We seemed to be embedded in a neural ganglion. I sat on a couch and Grof in a swivel chair. He rolled quite close to me and leaned forward as he spoke — fortunately, because he spoke softly. His eyes were dark and liquid — acid eyes, I thought. He listened to and answered my questions thoughtfully. And yet even when addressing the most esoteric and bizarre topics, he sounded oddly clinical, detached, remote, just as he did in his book *LSD Psychotherapy*.

Recounting his history, Grof told me that he entered medical school envisioning a career in Freudian-style psychiatry. He had joined a psychoanalytic discussion group while in his teens; later he underwent a seven-year psychoanalysis with the president of the Czech psychoanalytic association. But in the 1950s biochemical theories and therapies for mental illness were already beginning to challenge Freud's hegemony. The shift was precipitated by the discovery of drugs such as reserpine and chlorpromazine, which suppress psychosis, and LSD, which seemed to mimic psychosis. LSD's potency implied that mental illness might also be caused by subtle chemical changes in the brain. "Psychosis would basically not be a psychiatric problem but chemical, biochemical," Grof explained.

While still a medical student, Grof served as a "sitter" for subjects given LSD in an experiment supervised by one of his professors. He first took LSD himself in 1956, just after his graduation, as part of a study of how the drug affects brain wave patterns. Grof wore electrodes pasted to his scalp, and in the middle of his session a strobe light flashed in his face. "My consciousness was catapulted out of my body," Grof recalled. "I had no more boundaries. I became the universe. All kinds of cosmology, from the big bang to black holes, white holes." He eventually took high doses of LSD — ranging from 300 to 1,200 micrograms — more than a hundred times, and supervised more than four thousand LSD sessions involving others. Grof came to view LSD not as a psychotomimetic, a toxin that induces artificial psychosis, but as an amplifier of latent mental processes. "It's bringing out of the psyche stuff that's already there rather than an artifact," Grof said.

He concluded that conventional Freudian psychoanalysis could not account for the tremendous variety of experiences inspired by LSD. The drug often dredged up material from the unconscious — such as long-repressed memories from childhood and infancy — that Freudians could explain. But he and his subjects had visions that transcended their personal history. They were transformed into other human beings, other species, mythological figures. To make sense of LSD's transpersonal effects, Grof turned to the work of Freudian apostates such as Carl Jung, Wilhelm Reich, and Otto Rank, who saw more merit than Freud did in our spiritual yearnings and experiences. Grof also reached beyond psychology to Hinduism, Buddhism, and other religious traditions.

Grof dismissed two common interpretations of the LSD phenomenon. "Ultramaterialistic" scientists, he complained, have argued that LSD is the nail in the coffin of mysticism, since it implies that all these visions are just delusions produced by aberrant brain chemistry. Meanwhile, some philosophers and theologians have contended that drug-induced experiences are inferior to those produced by meditation, yoga, prayer, and other genuine spiritual practices. In other words, psychedelic experiences "might mimic spiritual experiences, but they are not really as valuable. They are pseudomystical, not mystical." (Steven Katz, whom I introduced in chapter two, espoused this view.)

His own psychedelic experiences had been "unquestionably spiritual," Grof said, and he knew many religious teachers from different traditions whose deepest spiritual experiences were drug-induced. That is not to say that LSD always produces a spiritual experience. "It's not something that takes you unconditionally, directly, to God." There are no universal physiological or psychological reactions to LSD, and the drug can amplify negative as well as positive human propensities. Ideally, mystical states help you "feel more connected and more part of things and more compassionate." But if the ego boundaries are only partially dissolved, the experience can lead to narcissism or worse. "The extreme of that is the psychotic who is convinced of his own divinity."

Some of Grof's experiences terrified him. He once became stuck in a birth canal with no apparent way out, and he had visions of concentration camps and other horrors. But these dark passages invariably gave way to sensations of rapture and resolution. "I never had a session which was very negative," he said. "That doesn't mean it can't happen. I've seen it happen to others."

Grof credited his drug sessions with helping him achieve a kind of spiritual equilibrium. "I've had experiences which were so difficult that it just permanently shifted the baseline," he said. "Everyday things just really don't get to you that much." Grof also came to terms with death. "Rightly or wrongly, I have the feeling I have been there. I have actually viewed death and thought I died already. I was surprised when I came back. So my present feeling is, it's going to be a fantastic trip. I don't have any sense that it's going to be the end of everything."

Reincarnation and Astrology

I'm not sure exactly when I realized just how, well, paranormal Grof's vision of reality is. Perhaps it was when we discussed his belief that our births have long-lasting effects on our psyches. Ordinarily, Grof said, we repress all memories of our birth, because "it is so painful we just don't want to deal with it." The psychoanalyst Otto Rank proposed a similar theory in the 1920s but did not produce much evidence to support it, and most neuroscientists now believe that the fetal and newborn brain cannot record lasting memories. But Grof claimed that Rank's theory has been amply corroborated by the perinatal visions of people in his LSD and holotropic breathwork sessions. These visions are not hallucinations but memories, Grof said.

He told me about an Austrian-born psychologist who was initially skeptical of the birth trauma theory but changed his mind after a holotropic session. The psychologist had had a powerful vision of his own birth, accompanied by a strong smell of leather. Afterward he called his mother, and without mentioning his session he asked her about his birth. She told him that she had been working in a shop that made lederhosen; her labor had come on so rapidly that she gave birth right there in the shop.

Grof asserted that during psychedelic and holotropic sessions we can recall not only our births but also our past lives. During his own LSD trips, he relived his previous incarnations with a clarity and richness of detail that could not possibly have been due solely to his imagination. "You can find yourself in another century, another culture, identified with another person," he said. "It goes way beyond anything that you knew intellectually."

Often these experiences of past lives provide insights into your current life. You may find that you have chronic pain in your shoulder because you were shot there when you were a soldier during the Civil War. Your asthma may be linked to your death from smoke inhalation in a previous life. You may realize that your troubled relationship with someone in this life reflects a conflict between your previous incarnations. These insights can have enormous therapeutic potential. "You suddenly have understanding why you have problems with this person, and why you have particular kinds of problems. Then you complete that experience and you have a sense of clearing. Then if you look, you can find out that at exactly the same time, something happened to the other person, and that person's point of view has also changed."

These are "facts," Grof assured me. "If somebody can explain it within a materialistic worldview, I will be the first one to be excited about it. But I haven't been able to do that." The only way to explain these facts, Grof said, is with a new scientific paradigm, which accepts that consciousness is not merely an epiphenomenon of the brain but can exist as an independent entity.

Grof derided the "ultramaterialists" who cling to the belief that the mind can be understood in purely physiological terms. He was particularly contemptuous of Francis Crick, who was a leading figure in the study of consciousness. He ridiculed Crick's proposal that consciousness can be understood by unraveling the neural mechanisms underlying vision, which is just one component of cognition. I briefly stuck up for scientific materialism, and for Crick, whom I had once interviewed. Crick, I suggested, would probably grant that he is looking at one piece of the puzzle; he would argue that consciousness is so complex that scientists can advance only by tackling small parts of it. That's fine, Grof replied, but then Crick should admit the limitations of his approach instead of implying that everyone who seeks a more expansive view is an "idiot." "We have all these observations that consciousness can operate without the body," Grof said. "But we still teach that consciousness somehow comes out of the brain."

Many ostensibly mainstream scientists, Grof said, are secretly sympathetic toward his transpersonal, "new paradigm" view. Since he is no longer part of the academic establishment, he can speak his mind without fear of harming his reputation. When he gives talks at universities, faculty members sometimes approach him afterward, to speak one on one. To his amusement, they invariably tell him that they agree with much of what he says but cannot say so publicly, because all their colleagues are so closed-minded.

My perusal of *LSD Psychotherapy* had left me with the impression that Grof was a psi agnostic, but now I saw he was a hard-core believer, a sheep. To convince me that gurus have psychic powers, which Hindus call *siddhis*, Grof told me a story about Muktananda, an Indian guru popular in the 1970s. Just before he first met Muktananda in India, Grof told his wife, Christina, about a psychedelic vision he had once had of the Hindu deity Shiva. When Grof was ushered into Muktananda's presence minutes later, the guru lifted up his trademark sunglasses, stared into Grof's eyes, and said: "I can tell you are a man who has seen Shiva. This is very good."*

Grof's transpersonal psychology was quite different from that of Ken Wilber, whose collected works Grof displayed in his office. Wilber expressed respect for mainstream, materialistic science. He claimed to have had moments of psychic communion with his late wife but asserted that

---

*Muktananda was the leader of a spiritual path called Siddha Yoga. Before he died in 1982, some of his followers accused him of sexually molesting young girls in his compound under the pretext of checking their virginity. (See Schwartz, *What Really Matters*, pp. 136–137; and "Crimes of the Soul," by Jill Newmark, Marian Jones, and Dennis Gersten, *Psychology Today*, March/April 1998, p. 55.) The daughter of a former associate of Muktananda's continues to oversee a popular offshoot of Siddha Yoga called SYDA Yoga.

mysticism has nothing to do with the paranormal. Although Wilber found reincarnation intellectually intriguing, he disavowed firsthand knowledge of past lives.

Grof, in contrast, was contemptuous of mainstream science, insisting that it must yield to a "new paradigm" that acknowledges mind's primacy over matter. He declared that his and others' transpersonal experiences have established reincarnation and other supernatural phenomena as "facts."* Grof exemplified the kind of thinking that Susan Blackmore, the mystical sheep turned goat, has shown underlies belief in psi. He saw no coincidences, only supernatural synchronicities. He seemed to believe in virtually all occult phenomena, including astrology. He claimed that a colleague, Richard Tarnas, had compiled evidence that planetary alignments affect our spiritual development and even the outcome of specific LSD and holotropic sessions.

Grof assured me that he would be willing to abandon his fundamental premise that minds can exist independently of bodies if someone could show him how astral projection and other out-of-body phenomena can be materialistically explained. Of course Blackmore, among others, has provided materialistic explanations of remote viewing, astral projection, and other supernatural phenomena. Blackmore has also examined Grof's proposal that people who have tunnel visions are reliving their passage through the birth canal. If this theory is true, she reasoned, then those delivered by cesarean section should not see tunnels during their near-death experiences. Yet she found that people delivered by cesarean section see tunnels as often as people delivered vaginally.

---

*Grof and other defenders of reincarnation often cite research done by Ian Stevenson, a psychiatrist and former professor at the University of Virginia who for decades has been studying cases in which people claim to remember past lives. Stevenson has presented his results in various books, notably the 2,268-page, two-volume *Reincarnation and Biology* (Praeger, Westport, Connecticut, 1997). One of the few skeptics to scrutinize Stevenson's voluminous work is Leonard Angel, a philosopher at Douglas College in Canada and a Zen Buddhist, who contends that even the best of Stevenson's cases fall apart upon close examination. Angel critiqued Stevenson's work in *Enlightenment East and West* (State University of New York Press, Albany, 1994, pp. 273–290), a book whose larger point is that mysticism is compatible with materialistic science; in "Empirical Evidence for Reincarnation?" (*Skeptical Inquirer,* vol. 18, no. 5, 1994, pp. 481–487); and in "Backwards Reasoning in Ian Stevenson's *Reincarnation and Biology*" (*Skeptic,* vol. 9, no. 2, 2001, pp. 72–78). For a journalistic account of Ian Stevenson's research by an editor for the *Washington Post,* see *Old Souls: The Scientific Evidence for Past Lives,* by Thomas Shroder (Simon and Schuster, New York, 1999).

Another critic of Grof's work is Franz Vollenweider. Based on his experience as a psychiatrist and an investigator of psychedelics, he does not believe that the trauma of birth has the kind of profound and persistent effect on our psyches that Grof postulates. Vollenweider, who has heard Grof lecture and has read his books, said that Grof struck him as someone "trapped within his own metaphysical system." Vollenweider added that Grof's decades-old research on the therapeutic potential of LSD was not carried out in the careful, controlled fashion that is necessary to demonstrate genuine benefits.

As for Grof's contention that in psychedelic and holotropic trances we see and learn things completely beyond our personal experience, an alternative hypothesis was presented decades ago by two other pioneering LSD researchers: Jean Houston, who later became a leader of the human potential movement, and her husband, the anthropologist Robert Masters. In the early 1960s Houston and Masters supervised more than two hundred LSD sessions and interviewed hundreds of other people who had taken psychedelics. They presented their results in 1966 in *The Varieties of Psychedelic Experience*, which remains a classic of psychedelic literature.

Houston and Masters could hardly be accused of being anti-drug; they asserted that psychedelics have great scientific, therapeutic, and spiritual potential. But they offered a somewhat deflating interpretation of the fantastical transpersonal visions reported by their subjects. The average American, they pointed out, "is exposed through his reading of newspapers, magazines, TV-watching, to enormous amounts of exoteric and esoteric information," which is "stored in regions of the mind accessible under certain conditions." The mythological, cosmic, historical, theological visions reported by many trippers, Houston and Masters concluded, "may constitute subliminal triumphs of *Time, Life, Newsweek*."

WHO AM I *REALLY?*

From a strictly rational standpoint, among the scientists I admire most are James Austin, Susan Blackmore, and Franz Vollenweider. They are skeptics, rationalists, empiricists who don't waste time fretting over unanswerable theological riddles such as the one I call the nirvana paradox: If nirvana is so great, why does God create? But as much as I admire their positivism, their eschewal of unconfirmable metaphysical speculation, I cannot resist that sort of speculation myself, and I find myself drawn to others who share my predilection.

I suppose that was why I found Grof hard to dismiss. He defended even outlandish notions with such gravity and authority that — in his presence, at least — I found myself questioning my biases rather than his. Maybe I should look at reincarnation more closely, I thought, and telepathy, too. Now and then I felt almost mesmerized. When Grof told me that he, Grof, and I, Horgan, are "part of the same unified web," my flesh prickled as he transfixed me with his dark, luminous eyes. I remembered how, in the aftermath of my 1981 drug trip, I felt my connection to others so intensely that I could not look anyone in the eyes.

That psychedelic episode also made me sympathetic toward Grof's transpersonal cosmology. I found myself transformed into people and things totally foreign to my personal history: a double helix, an apeman squatting on the veldt, a superintelligent cosmic computer. The visions possessed a mythical, archetypal quality that my dreams lack. They seemed not absurd and meaningless, like many of my dreams, but almost *too* meaningful. Afterward I wondered, Where did *those* come from?

As Jean Houston and Robert Masters suggested in *The Varieties of Psychedelic Experience*, my hallucinations may have been constructed out of all the cultural bric-a-brac, or memes, I'd absorbed after decades of reading, watching movies and television, and so on. But my visions seemed too vivid, finely detailed, and artful — too laden with metaphorical and metaphysical significance — to be the products of my puny, personal brain. I was tempted to agree with Grof that I had actually left my individual mind behind and traveled into another, more expansive, transpersonal realm.

I was also impressed by his boldness in addressing ancient theological riddles. He was wisely less dogmatic when offering these speculations than when defending ESP and reincarnation. When I asked if transpersonal experiences can solve the mystery of existence, revealing how and why the universe was created, Grof shook his head. "Not *solve*," he replied. "Because ultimately we are stuck with a mystery that you cannot account for." But our transpersonal revelations can suggest possible answers to some theological questions.

The transmutations of the self wrought by LSD, Grof said, raise a profound metaphysical question: "If I can be so many different things, then who am I *really?*" Hinduism provides as good an answer as any: All things are manifestations of one thing, the cosmic mind, the creative principle, which Hindus call Brahman. The next question is, Why does creation occur at all? Virtually all spiritual systems, Grof pointed out, teach that this mundane reality, in which we live as distinct, mortal individuals, is flawed. "This is a garbage level, the quagmire of death-rebirth, the valley of tears." According to Hinduism and Buddhism — and, to a lesser extent, the Western faiths — our ultimate goal should be transcending our separateness and becoming one with our divine source.

But if mystical unity is all it's cracked up to be, Grof asked, why does creation occur at all? Why doesn't the creative source of all things remain in its perfect, unified state? Evidently, nirvana is not really that self-fulfilling. Just as we, the individual products of creation, are compelled to reunite with our divine source, so that source is compelled to divide itself again. "You have this peculiar situation," Grof said. "You have this source that has this propensity to create and project itself out. Then, when you end up here as a separate unit, there is trouble, suffering, and so on," all of which propel us back toward the source. Grof calls these two opposed creative forces *hylotropic* and *holotropic.* The former is centrifugal, matter-oriented, pushing toward fragmentation and multiplicity; the latter is centripetal, spiritual, pushing toward wholeness and unity.

Sifting through his and others' transpersonal experiences and through countless mystical texts, Grof has sought, in effect, to psychoanalyze God, to understand the motives underlying divine creation. One possibility is that God is like an artistic genius who creates out of sheer joy, or *lila,* the Hindu term for divine playfulness. But God may have darker motives. He may create out of loneliness, boredom, anxiety, a craving for self-knowledge or adulation, or some other deficiency or need. Grof cautioned that human loneliness and boredom, our desire for love and self-awareness are only faint echoes of their divine counterparts. "Those are dimensions of experience that you can't imagine in an ordinary state of consciousness."

Grof then took on the old theological conundrum: Why would God create a world with so much injustice and suffering? The Indian swami Ramakrishna once said evil is needed to "thicken the plot" — in other words, to make existence interesting. This notion offended Grof at first — "We are talking about children starving in Ethiopia, the Holocaust" — but gradually it made sense to him. Without famine, war, droughts, racism, totalitarian regimes, religious fanaticism, and other blights, human history would be too dull. Without disease, we would never have invented medicine. Without slavery and repression, we would have no liberation movements. "You lose a big part of the cosmic drama," Grof said. "You end up with a plot so flat that, if it were made into a movie, we wouldn't go." In other words, God must make creation as interesting as possible to distract Himself from His existential plight. Hence evil.

Listening to Grof, I recalled the squabble I had witnessed that morning: Mom and Dad and the two boys screaming at one another, furious and miserable. The thicken-the-plot concept implied that to some extent these conflicts are necessary. If love, the urge to merge, were unopposed by any repulsive forces, the result would be as catastrophic as if gravity were the only force in the universe. We would collapse into a black hole. We fight, break apart, suffer, because we must, for all eternity. Could that be right?

As chilling as Grof's theodicy was, it had much in common with my "gnosticism lite" hypothesis and with the theological propositions of others. Ken Wilber suggested that God divides Himself into many selves who have forgotten their true nature, because "no one likes to eat dinner alone." Gnostics, Kabalists, Christian theologians, and others who have obsessed over the problem of natural evil have shared Grof's suspicion that there must be something wrong with God.*

The thicken-the-plot idea has also been echoed by Freeman Dyson, one of the architects of particle physics. In *Infinite in All Directions,* Dyson pondered why there is so much violence, hardship, and evil in the world. He suggested that reality might be ruled by what he calls "the principle of maximum diversity." This principle, he explained,

> operates at both the physical and the mental level. It says that the laws of nature and the initial conditions are such as to make the universe

---

*So-called process theologians, notably Alfred North Whitehead (1861–1947), argued that God cannot be a perfect, static essence but must grow, evolve, and suffer just as humans do. For a recent elaboration of this view, see *The Creative Suffering of God,* by Paul Fiddes (Oxford University Press, Oxford, 1992). Yehudah Bauer, an Israeli authority on the Holocaust, has said that God must be either Satan or a "nebbish" (as quoted in *Explaining Hitler,* by Ron Rosenbaum [HarperPerennial, New York, 1998, p. 283]). Similarly, the poet Annie Dillard has referred to "God the semi-competent" (as quoted in *Why Religion Matters,* by Huston Smith [HarperCollins, New York, 2001]). A vision of God being driven insane by His awesome responsibility can be found in *The Politics of Experience,* by R. D. Laing (Ballantine Books, New York, 1967, p. 158).

> as interesting as possible. As a result, life is possible but not too easy. Always when things are dull, something turns up to challenge us and to stop us from settling into a rut. Examples of things which made life difficult are all around us: comet impacts, ice ages, weapons, plagues, nuclear fission, computers, sex, sin, and death. Not all challenges can be overcome, and so we have tragedy. Maximum diversity often leads to maximum stress. In the end we survive, but only by the skin of our teeth.

Like Grof, Dyson proposed that hardship, suffering, pain, chaos, conflict, and above all diversity are necessary, to make life interesting, to "thicken the plot." The principle of maximum diversity rules out the possibility that someday we will achieve a kind of utopia or heaven on earth in which all our troubles vanish. We will never attain what the Tibetan scholar Robert Thurman calls the Buddhaverse, a world in which we are all enlightened. Drama is more important than resolution. Life is, and must always be, a struggle.

If my meeting with Grof aroused my hopes that mysticism could solve age-old theological riddles, it also reminded me — as if I needed reminding — of the threat that mysticism poses to sanity. Just before I left Grof's house, Christina appeared at the door of his office, slightly breathless. I had read a bit about her. In the 1970s, while she was practicing yoga and other spiritual disciplines, she began having transpersonal visions so powerful that she feared she was losing her mind. In 1975, after hearing about Grof, she sought him out for help in dealing with these states. The two fell in love and married, each for the second time, but Christina's troubles persisted. She began drinking heavily, and she suffered a nervous breakdown in 1986.

Then she pulled her life together. She wrote articles and books that drew lessons from her difficult personal history. Together with her husband, she argued that many psychological problems — among them depression, alcoholism, and drug addiction — are actually "spiritual emergencies" that stem from a deep-rooted yearning for spiritual meaning and consolation. Properly treated, she said, these spiritual crises offer a tremendous opportunity for growth.

After Christina and I exchanged greetings, she wrapped her arm around her husband and patted him on the back. "He's a good one, this one," she said as he maintained an inscrutable expression. I couldn't help scanning her for signs of some abiding woundedness, but all I saw was a tall, blond-haired woman with an incandescent smile and pale blue eyes so wide open she seemed almost startled. For some of us, mysticism represents not salvation but that from which we need to be saved.

DMT ALIENS IN HYPERSPACE

As far as Stanislav Grof has strayed from mainstream science, he remains tethered, albeit loosely, to conventional spirituality, such as Hinduism, and conventional psychology (if psychoanalysis and Jungianism can still be considered conventional). He has his limits. In *Psychology of the Future,* he noted that he had encountered many subjects over the years who claimed to have had encounters with aliens either in psychedelic sessions or under other circumstances. Grof found it "implausible" that these visions involved "actual visits of aliens from other worlds." He preferred the explanation that Jung set forth in an essay on "flying saucer" phenomena, that extraterrestrials represent "manifestations of archetypal elements from the collective unconscious."

Another psychonaut who grappled with the alien issue was John Lilly, pioneer of dolphin-language studies, inventor of sensory-isolation tanks, and scientific polymath. Lilly was the role model for the brilliant but unstable researcher played by William Hurt in the movie *Altered States.* (Lilly's career also inspired the earlier movie *Day of the Dolphin.*) He took massive amounts of LSD and the drug ketamine, which he called vitamin K, often while immersed in a sensory-isolation tank. (Ketamine is the drug with which the Russian psychiatrist Evgeny Krupitsky has treated alcoholics and drug addicts.) He claimed that in this state he entered a dimensionless hyperspace where he was approached by superintelligent, nonhuman beings who communicated with him telepathically. But like Stanislav Grof, Lilly seemed inclined to think that the beings were probably products of his own unconscious.*

*Lilly, who died on September 30, 2001, told the story of his adventuresome life in *The Center of the Cyclone* (Bantam Books, New York, 1972); and *The Scientist* (Ronin Publishing, Berkeley, California, 1988). See also the entertaining profile of the ketamine-addled but still formidable Lilly in *The Three-Pound Universe,* by Judith Hooper and Dick Teresi (Jeremy Tarcher, Los Angeles, 1986, pp. 271–280). Lilly told Hooper and Teresi that during certain transpersonal

A more radical interpretation of the "close encounter" phenomenon was recently proposed by Rick Strassman, a psychiatrist at the University of New Mexico Medical School. From 1990 to 1995, Strassman injected the drug dimethyltryptamine, or DMT, more than four hundred times into sixty volunteers; it was the first officially sanctioned test of psychedelics' effects on humans in the United States since the early 1970s. What makes DMT unique among the psychedelics is that it was detected in human blood in 1965 and in human brain tissue in 1972. These discoveries fueled speculation that endogenous DMT contributes to schizophrenia and other mental illnesses. By the late 1970s, investigations of DMT's role in psychosis were largely abandoned, along with most other psychedelic research involving humans.

Strassman obtained permission for his study, which he described in his 2001 book *DMT: The Spirit Molecule,* by arguing that the DMT theory of psychosis merited further investigation. As a Zen Buddhist, Strassman was also intrigued by the possibility that endogenous DMT plays a role in triggering mystical experiences.* To an extent, the DMT sessions fulfilled Strassman's expectations. Many of the subjects reported classic mystical sensations: bliss, ineffability, timelessness, and certainty that consciousness continues after the death of the body. Others underwent stereotypical near-death experiences; they felt themselves leaving their bodies and moving through a tunnel toward a radiant light.

But to Strassman's surprise, nearly half of his sixty subjects encountered bizarre, otherworldly beings, described as clowns, elves, robots, insects, E.T.-style humanoids, or uncategorizable "entities." These entities were not always benign. Some expressed indifference or hostility toward the volunteers or performed painful experiments on them. One man was

---

episodes, you experience your divinity so completely that "you must *forget* about them when you come back. You must forget you're omnipotent and omniscient and take the game seriously, so you'll have sex, beget children, and the whole human scenario" (p. 277). Lilly's warning evokes what Ken Wilber calls amnesis, or divine forgetfulness, and what the philosopher Alan Watts called "the taboo against knowing who you are."

*In 1996 Strassman described his DMT research in a Buddhist magazine: "Sitting for Sessions: Dharma and DMT Research" (*Tricycle,* Fall 1996, pp. 81–88). He admitted that some members of a Zen community to which he belonged had condemned his research, which they felt would associate Buddhism with illicit drugs. In the same issue *Tricycle* reported that 83 percent of a group of 1,454 readers and Web-site visitors admitted having used psychedelics at least once; 24 percent were currently taking psychedelics.

eaten alive by insectoid creatures, and another was anally raped by two reptilian monsters. Some volunteers were understandably terrified during their sessions and remained shaken afterward.

Strassman, who like Stanislav Grof was trained in Freudian psychoanalysis, initially interpreted these entities as embodiments of volunteers' subconscious fears or desires. When the volunteers insisted on the reality of their visions, Strassman eventually proposed a more radical hypothesis: the DMT entities dwell in one of the multidimensional hyperspaces postulated by certain highly speculative theories of physics. Normally we have no access to these other realms, but like a rocket ship, DMT somehow propels us into one of these hyperspace realities, where we encounter the alien beings. Strassman was proposing, in other words, that the robotic, clownish, insectoid entities are not phantasms of our overstimulated human brains; they exist somewhere *out there.*

For some reason, I find it easier to ponder the possibility of a neurotic God or Demiurge than of aliens living in hyperspace. Instead of helping us in our search for a reasonable, comforting theology, these aliens, it seems to me, complicate matters.* As the physicist I. I. Rabi reportedly blurted out, when confronted with evidence of a totally unexpected particle, the muon, in the 1950s, "Who ordered *that?*" But I have more substantive objections to those hyperspace aliens. The theories of physics cited by Strassman in support of his assertion are not only untested but, according to many physicists, probably untestable, at least in the foreseeable future; they are little more than science fiction in mathematical form. Moreover, hallucinations of strange beings can be triggered not only by psychedelics but also by other conditions.

Susan Blackmore notes that many cultures have purveyed legends about demonic beings who torment us while we lie helplessly in bed at night. These creatures were called succubi and incubi in medieval Eu-

*In July 2001, Rick Strassman told me, in an exchange of e-mails, that he still believed that DMT could be spiritually beneficial. When I asked how anyone would benefit spiritually by contacting the DMT entities, Strassman replied that he had been mulling over that question himself: "Even if these 'beings' were 'real,' are they of any real value to us? What do they have to offer that we don't already know, but just are not putting into practice? Maybe, though, it might be a case of people doing things that celebrities do and say, but because of the source! If these beings 'teach' peace and the like, maybe it will make us more likely to practice it." So the DMT entities, I asked, would be like alien Richard Geres, whom we listen to because of their supposedly exalted status? "Sort of like that," Strassman responded.

rope, the Old Hag in Newfoundland, and Popobawa in Zanzibar. Similarly, present-day alien abductees often report being approached while they are in bed. Blackmore suspects that these nightmarish visitations are caused by a condition called sleep paralysis. When you dream, your ability to control your muscles shuts down, so you do not act out your dreams. Occasionally you may partially awaken, becoming dimly aware that you are lying in bed while still subject to sleep paralysis and hallucinations. In this state, Blackmore explained in *The Meme Machine,* you may perceive odd noises, lights, and vibrations, and you may have "a powerful sense that there is somebody or something in the room with you."

People may perceive "beings" while fully awake as a result of disruption of the communication between the brain's hemispheres, according to Michael Persinger. As he told me during my visit to Laurentian University, many brain-damaged patients are plagued by visions or intuitions of "sensed presences." These sensations can also be caused by severe stress, bereavement, lack of sleep, and other factors.

A neurological disorder called Charles Bonnet syndrome, usually caused by damage to the eyes or some other part of the visual system, also brings on hallucinations of animate and inanimate objects, often in a sketchy, cartoonish form. The apparitions may be mundane, but they can also be exotic. In one study, more than one in ten elderly people with eye disease reported seeing ghosts, dragons, angels, circus animals, clowns, and elves. "Given how common this syndrome is," the neuroscientist V. S. Ramachandran wrote in his book *Phantoms in the Brain,* "I am tempted to wonder whether the occasional reports of 'true' sightings of ghosts, UFO's, and angels by otherwise sane, intelligent people may be merely examples of Charles Bonnet hallucinations."

I am tempted to wonder the same thing. But if anyone could persuade me to entertain the notion of hyperspace aliens, it would be Terence McKenna.

10

The Man in the Purple Sparkly Suit

In May 1999, I filed into an auditorium in New York City along with several hundred people, young and old, sporting gray ponytails or blue mohawks or orange spikes or shaved, tatooed skulls. We were gathering to see the psychedelic author Terence McKenna perform one of his free-form talks, or "raves." Given the anarchic nature of the psychedelic subculture, terms like *psychedelic leader* or *psychedelic spokesman* have an oxymoronic ring to them, but if anyone deserves those designations, McKenna does. The incorrigible trickster Timothy Leary himself, just before he died, dubbed McKenna "the Timothy Leary of the 90's."

Like Leary, McKenna is less a scientist or philosopher than a poet, performance artist, jester, shaman. He does not try, as Huston Smith and Stanislav Grof do, to make the psychedelic experience conform to the perennial philosophy or one of its institutional embodiments, whether it be Hinduism or Islam. McKenna extols psychedelics as a path distinct from and superior to that of any mainstream religion. (He disdains the term *entheogen*, calling it "a clumsy word freighted with theological baggage.")

He is particularly fond of DMT, the drug given to volunteers by Rick Strassman in the early 1990s. As Strassman has acknowledged, well before his book *DMT: The Spirit Molecule* was published, McKenna was drawing attention to DMT's special status as an endogenous human psychedelic and describing the entities, or "self-transforming machine elves," encountered under its influence. Like Strassman, McKenna speculated that DMT plunges us not into our own unconscious but into some sort of hyperspace or "interdimensional nexus" with its own alien inhabitants. In other words, yes, the entities are *out there.*

Raised in a Colorado ranching town, McKenna discovered psychedelics in 1965 when he enrolled at the University of California at Berkeley, from which he eventually earned a degree in ecology, resource recovery, and shamanism (ah, those were the days). He became an itinerant scholar-adventurer, traveling to the Far East and South America in search of exotic mind-altering philosophies and substances, including two of his favorites, psilocybin mushrooms and ayahuasca. McKenna first attracted widespread attention in 1975 after the publication of *The Invisible Landscape*, a book of wild psychedelic speculation, cowritten with his brother, Dennis.

McKenna can play the serious scholar when he chooses. His 1992 book *The Food of the Gods* is a well-researched argument — complete with footnotes and bibliography — that mind-expanding flora and fungi catalyzed the transformation of our brutish ancestors into cultured modern humans. The visions produced by plants and mushrooms containing psilocybin, DMT, and other psychedelics were the seeds from which language grew, followed by the arts, religion, philosophy, science, and all of human culture, McKenna asserted. By outlawing psychedelics, we have cut ourselves off from the wellspring of our humanity.* He pointed to archaeological evidence that psychedelics played a role in the origins of art and religion: Some of the earliest known art — millennia-old petroglyphs found in Australia, Africa, the Americas, and elsewhere — depicts the distinctive geometric hallucinations, or form constants, triggered by psychedelics. Form constants recur in a wide range of ancient sacred art, from the mandalas of Tibetan Buddhists to the yarn paintings of Huichol Indians.

The writers of the Indian Vedas — sacred Hindu texts written more than three thousand years ago — claimed to receive inspiration from the

---

*In support of this thesis, McKenna cited *Intoxication* (E. P. Dutton, New York, 1989), in which Ronald Siegel, a psychopharmacologist at UCLA, contended that the desire for intoxication is "the fourth drive," after hunger, thirst, and sex. This compulsion has deep evolutionary roots; field biologists have observed chimpanzees, elephants, parrots, and other species deliberately ingesting fermented fruit and other intoxicants. Similar arguments were advanced almost two decades earlier by Andrew Weil in *The Natural Mind* (Houghton Mifflin, Boston, 1998, originally published in 1972). Weil, a Harvard-trained physician who became a leading advocate of alternative medicine, asserted that intoxication "has been a feature of human life in all places on the earth and in all ages of history" (p. 17) and stems from "an innate, normal drive analogous to hunger or the sexual drive" (p. 19). See also "Psychotropic Substance Seeking: Evolutionary Pathology or Adaptation?," by R. J. Sullivan and E. H. Hagen (*Addiction*, vol. 97, 2000, pp. 389–400), in which two anthropologists argue that our fondness for stimulants has deep evolutionary roots and originally had adaptive value.

mysterious potion soma. McKenna argued that soma was probably made from one of the psilocybin species (other scholars have favored the mushroom species *Amanita muscaria,* also called the fly agaric). He suspected that entheogenic brews were used in the Eleusinian Mysteries, sacred celebrations that took place each fall in ancient Greece for more than a thousand years. These rites may have given impetus to the philosophies of Plato and other seminal Greek thinkers.

Psychotropic drugs, McKenna added, played an especially vital role in shamanism, the ur-religion that predates all modern ones. He defined shamanism as "the tradition of healing, divination, and theatrical performance based on natural magic" that emerged among diverse groups of humans during the paleolithic era. By means of what the scholar of shamanism Mircea Eliade called the "archaic techniques of ecstasy," including not only drugs but also chanting, dance, fasting, and other mind-altering methods, the shaman learns how to "travel in the spirit realm at will."*

Other scholars have argued that psychotropic substances and shamanism made a crucial contribution to the evolution of culture. What sets McKenna apart is his conflation of the archaic, shamanic worldview with the latest scientific developments, such as fractals, artificial intelligence, and the search for extraterrestrial intelligence. At the same time, he suggests that shamanism has more in common with postmodernism than it does with either science or religion. Like postmodernism, shamanism implies that we do not so much discover reality as invent or imagine it. "For

---

*The Romanian scholar Mircea Eliade helped to popularize the notion of shamanism as a distinct, worldwide form of spirituality with his book *Shamanism: Archaic Techniques of Ecstasy* (Pantheon, New York, 1964). Eliade was also an anti-Semite and a fascist, according to "Witness to Fascism," by Peter Gay (*New York Review of Books,* October 4, 2001, pp. 44–47). Another controversial popularizer of shamanism was the anthropologist Carlos Castaneda, author of *The Teachings of Don Juan* (University of California Press, Berkeley, 1968). In this book and its many sequels, Castaneda described his apprenticeship with a Mexican shaman named Don Juan. Other scholars have long maintained that Castaneda's books were largely if not entirely fictitious. The reclusive Castaneda, who died on April 27, 1998, never defended himself against these charges. For reviews of the controversy over Castaneda's books, see *The Don Juan Papers,* edited by Richard de Mille (Wadsworth Publishing, Belmont, California, 1980) and "Mystery Man's Death Can't End the Mystery," by Peter Applebome (*New York Times,* August 19, 1998, p. E1). For a defense of neo-shamanism, particulary aimed at criticisms leveled against it by Ken Wilber, see "The Epistemology and Technologies of Shamanic States of Consciousness," by Stanley Krippner (*Journal of Consciousness Studies,* vol. 7, no. 11–12, 2000, pp. 93–118).

the shaman, the cosmos is a tale that becomes true as it is told and as it tells itself," McKenna wrote.

McKenna himself excels at telling tales. In *True Hallucinations*, an autobiographical work published in 1993, his ironic, pseudoscholarly style evokes that of the Argentinian fabulist Jorge Luis Borges. So does the plot, in which an ostensibly rational narrator struggles to maintain his sanity as he confronts occult mysteries. The tone is established in the subtitle ("Being an Account of the Author's Extraordinary Adventures in the Devil's Paradise") and chapter headings ("Looking Backward: In which several miracles are recounted, not the least of which is the appearance of James and Nora Joyce disguised as poultry"). The book's main narrative follows Terence McKenna, his brother Dennis, and several companions on a journey into Amazonia in the early 1970s in search of shamanic knowledge. Settling for several weeks in the tiny Colombian hamlet of La Chorerra, they ingested enormous quantities of marijuana and psilocybin mushrooms, and they prepared and consumed a batch of ayahuasca.

Much to the alarm of Terence and his comrades, Dennis's behavior became increasingly erratic, and he finally lapsed into a raving unintelligibility that lasted for weeks. Terence's condition wasn't much better. Coincidences vanished; even shifts in the weather were messages intended for him. He became convinced that Dennis was channeling the spirit of their mother, who had died four months earlier, and he saw flying saucers. He began to suspect that the tryptamines — including DMT and psilocybin, the primary active ingredients of ayahuasca and magic mushrooms, respectively — might represent messages or spores sent from an alien civilization, or "overmind."

This journey also inspired McKenna's timewave theory, which holds that existence springs from the interaction between two opposing forces, one conservative and the other creative, or novelty-generating. In the 1980s, McKenna devised a mathematical model that charts the ebbs and surges of novelty — wars, revolutions, famines, plagues, scientific and technological advances — throughout human history. When he extrapolated the model into the future, it predicted a huge spike in novelty on December 22, 2012.

The ultra-hip archness of *True Hallucinations* now and then gives way to moments of genuine poignancy. McKenna confessed that as he was finishing the book, his sixteen-year marriage to his wife Kat, with whom he'd had two children, was dissolving. All his psychedelic insights, he said, had "done nothing to mitigate or ward off the ordinary vicissitudes of life. Like the Soul in Yeats's poem I am still an eternal thing fastened to the body of a dying animal." This mixture of earnestness and irony is the key to McKenna's charm. He seems unsure himself whether his psychedelic visions are true or just delusions, whether he is serious when he talks about the timewave and DMT elves and the impending apocalypse or just goofing.

Hoping to get a better sense of McKenna's true beliefs, I arranged to meet him in New York. He had flown there from his home in Hawaii to give a talk. On the night of his performance, as I took my tape recorder and pad out of my backpack, a man in front of me turned around to chat. He was bearded, about my age, a bit wild-eyed. I'll call him Glen. He had never seen "Terence" before, but he had all of Terence's books and two of his taped raves. He was blown away by the prediction that the shit was going to hit the fan in 2012. He hoped that tonight Terence would get a little more specific about what was going to happen. Glen had dropped a lot of acid when he was young. In fact, he liked drugs so much that he became a pharmacist. Working in a pharmacy got boring after a while, so he became a psychotherapist specializing in "energy healing." Glen hadn't done acid in a long time. He really wanted to do some again, but he didn't know where to get it. Glen looked at me expectantly. When I shrugged, he glumly turned his back to me.

Applause erupted as McKenna strode onto the stage and sat in a spotlit armchair. Tall, bearded, owlish, with curly brown hair going gray, he looked like a gaunt, Celtic Allen Ginsberg. His voice had a wheedling, nasal edge. His speech was much like his prose, a bouillabaisse of scientific and high-tech visions, primordial lore, environmental fear-mongering, anarchic political commentary, and psychedelic woolgathering.

He kept wandering down digressive paths that somehow circled back to his main theme: The purpose of existence is novelty-generation, and our culture is generating novelty at an accelerating rate. The Internet, nanotechnology, genetic engineering, artificial intelligence, all are shattering our old paradigms and bearing us toward some fantastic, psychedelic future. And according to his calculations, on December 22, 2012, something big is going to happen, something apocalyptic that will bring about "the end of life as we know it."

At the moment McKenna was leaning toward something related to artificial intelligence, which he believed is well on its way to producing machines infinitely smarter than their human creators. "If the hyperintelligent AI is not filled with bodhisattva compassion," he warned, "then our ass is probably grass." McKenna presented all of this with a mischievous grin, daring us to take him seriously.

Raving at the Millennium

I was still unsure what to make of McKenna when I met him the next day for lunch atop the Millennium Hotel, a gleaming ebony monolith in lower Manhattan near the World Trade Center (then still standing). We sat in a booth beside a window, McKenna with his back to the glass, beyond which skyscrapers loomed. He looked even more owlish up close than he had onstage. He wore a black T-shirt emblazoned with a bronze dancing figure. Aztec? Mayan? The gap between his front teeth enhanced his air of raffishness.

When I told him that I had found his talk the previous evening entertaining, he rolled my descriptor slowly around in his mouth — "En-ter-tain-ing" — as if he weren't sure he liked its taste. I added that I thought he exaggerated the extent to which old scientific paradigms were crumbling and yielding to a radical new "psychedelic" vision of reality. I admitted that I had once worked for *Scientific American,* and I still shared that magazine's skeptical, conservative perspective.

Scientific American, McKenna assured me, is one of his favorite magazines. It is "incredibly psychedelic" and a major source of his inspiration. He had just read an article in *Scientific American* about a hypothetical cosmic force with "surreal" implications. "Essentially, what it says is that Newtonian spacetime is in the act of boiling away, and what will be left in only a couple or three billion years — if you calculate these processes fully — is a universe entirely defined by nonlocal quantum activity. That's about as woowoo as . . ."

A beetle-browed waiter appeared and asked, with an eastern European accent, if we would like drinks. McKenna ordered a vodka gimlet. A few minutes later, the waiter returned and said to McKenna, You ordered vodka tonic, yes? No, vodka *gimlet,* McKenna replied. Gym-let? the waiter repeated shakily. McKenna patiently explained how to make a vodka gimlet: vodka, lime juice, ice. The waiter, looking more perplexed, departed as

McKenna, mouth agape, watched him. "The concept of a gimlet is slipping away at the center of New York City? The apocalypse has struck!" His laughter had a wicked, Snidely Whiplash flair.

Anyone who reads *Scientific American,* McKenna continued, can see that science is in the throes of "an enormous crisis, or maybe not crisis but turmoil, based on the breakdown of paradigms." Just look at superstrings, parallel universes, hyperdimensions, time travel, and other bizarre notions emerging from physics. Then there are technological advances such as AI, which is on the verge of creating machines with superhuman intelligence. "Nobody knows what mind is," McKenna said, "operating at multigigahertz speeds in virtual realities unconstrained by gravity and economy of any sort."

I told McKenna that many of the supposed revolutionary advances he had read about in *Scientific American* and elsewhere were grossly oversold. Artificial intelligence, far from being on the verge of creating "hyperintelligent" machines, is a joke, a failure, with numerous failed prophecies behind it. As for superstrings, they're so small that it would take a particle accelerator the size of the Milky Way to detect them. That's why many physicists believe that superstring theory and its variants will turn out to be a theoretical dead end.

McKenna shrugged. Whether or not superstring theory pans out, no one can deny that physics is "a field in rich ferment, in need of broad new paradigmatic visions. I mean, sort of where physics was circa 1898, when no one could make sense of Planck's black-body studies and Einstein was chasing girls around bars in Zurich, and it just hadn't quite frozen out to be what it was." What was most impressive about McKenna's riffs was their apparent effortlessness.

The waiter arrived with our lunches. Fresh pepper? he asked, brandishing a howitzer-sized grinder. As the waiter ground away, I asked McKenna if he seriously believed that psilocybin mushrooms represent messages from an alien intelligence. His proposal isn't as whimsical as it sounds, he responded. Mushroom spores can survive the cold of outer space; in fact, mushroom cultivators here on earth store spores in liquid nitrogen. "So if somebody were designing a bioinformational package, a spore is how you would go. Millions of them pushed around by light pressure and gravitational dynamics would percolate throughout the galaxy."

Psilocybin's unusual chemical structure suggests an unusual origin,

McKenna added. It is "the only phosphoric acid–containing indole in all of nature," which indicates "that maybe it came from outside the terrestrial ecosystem." The personality of the mushroom, as revealed by the experiences it induces in humans, also has a science-fiction quality. "It presents itself as this particular slice of alien, aesthetic motif from Hollywood — the shiny metallic surfaces, the mercuroid forms, the piercing, instantaneous biointelligence." McKenna was enjoying himself; he uttered "aesthetic motif" and "mercuroid forms" with a kind of tactile relish.

I said my impression was that he was often . . . kidding. McKenna guffawed. "I'm Irish! What's your excuse!" He added soberly, "I'm cynical, and a fair cynic must also be cynical about himself." He called himself a "visionary fool" who "propounds this thing which is a trillion-to-one shot" — the timewave theory — and then "gets to live out the inevitably humorous implications of that."

He recognized that some people think he is "soft-headed" because of his rants about "self-transforming machine elves from hyperspace and all that." Actually, he had a "keen nose for other people's bullshit." He derided the claim — whose most prominent proponent is the Harvard psychiatrist John Mack — that aliens are abducting humans on a massive scale and molesting them with anal probes. McKenna liked to call these hypothetical aliens "pro bono proctologists from a nearby star system." He also despised New Age authors who claim to be channeling the spirits of long-dead sages. "These things are like intellectual viruses loose in the theater of discourse," he said. "And you can't really argue with these people, because they don't understand the rules of argument." McKenna added that "the howling tide of unreason beats against pure fact with incredible fury."

But I bet many of the people in your audience last night . . .

"Were of that ilk?"

Yes, I replied. I told McKenna that at his talk someone had handed me a leaflet advertising a lecture by John Mack on alien encounters. I said that McKenna's ruminations about extraterrestrial psilocybin spores and the coming apocalypse struck me as intellectual performance art, not to be taken too seriously, but I suspected that many of his fans take his ideas literally. I recalled my discussion with Glen, the man sitting in front of me, who had seemed obsessed with McKenna's apocalyptic prophecy.

McKenna nodded ruefully. "My function is largely pedagogical," he said, "trying to teach people, first of all, that the world is a weird, weird place. And then, so what do you do about it? Do you become a Scientologist? Do you return to your Irish Catholic roots? What is the response," he said, "to the discovery that the world really is totally weird?"

When I told him that his writing sometimes reminded me of Jorge Luis Borges's, McKenna was delighted; he was a Borges fan, too. He shared Borges's intuition that "scattered through the ordinary world there are books and artifacts and perhaps people who are like doorways into impossible realms, of impossible and contradictory truth, DMT being the chief example of this." DMT transports us not to heaven or hell but to a "parallel dimension that is somehow intimately tangled up with our own language processes and how we make reality." Modern science "operates on the assumption that there are no miracles at the macrophysical level. Well, I would put it to you, DMT *is* a miracle at the macrophysical level. And the smarter you are, the more impressive it is."

Up to this point, McKenna had absorbed all my questions with good humor. He seemed taken aback only when I mentioned that his brother, Dennis, had told me he suspected Terence had stopped taking drugs. McKenna scrutinized me through slitted eyes. "Maybe he thought you were the heat," he snarled. He assured me that he smokes a *lot* of cannabis, and he takes stronger substances occasionally, although not as often as he once did.

As if to assure me that he had not gone soft, McKenna mounted a spirited defense of the *via psychedelica*. Unlike Buddhism, Hinduism, and other mystical traditions, which are "*very* hierarchical," psychedelics offer direct, unmediated revelation, which is "so boundary-dissolving that it challenges all forms of social hierarchy." Psychedelics "are democratic," McKenna said. "They work for Joe Ordinary. And I *am* Joe Ordinary. I can't go and sweep up around the ashram for eighteen years or some rigmarole like that."

McKenna doubted a tale told in the counterculture classic *Be Here Now* by Baba Ram Dass, who, when he was a psychologist at Harvard in the early 1960s, was called Richard Alpert. Ram Dass recalled that when he gave an Indian guru a huge dose of LSD, the guru was unaffected because he already had such a profoundly mystical outlook. The implication was that spiritual practices such as meditation and yoga can induce the same powerful mystical states as drugs, but in a more stable, permanent fashion. McKenna suspected that Ram Dass's guru secretly palmed the LSD instead of consuming it; the swamis McKenna had met in his travels in the East were certainly capable of such a trick.*

McKenna granted that psychedelics do not necessarily promote moral behavior. "It has not produced a steady stream of canonizable human beings," he said drily. On the other hand, the same could be said of nonpsychedelic spiritual paths. After all, Hinduism and Buddhism have recently spewed forth "an endless stream of drunken, philandering, embezzling, skirt-chasing, cavorting roshis, rishis, geysheys, gurus. So I guess we can't judge a field of human endeavor by the moral behavior of its adherents, because all would be dismissed."

"I love humanity," McKenna added, "but I think we're closer to carnivorous monkeys than anything else." The waiter appeared to clear plates and ask if we wanted coffee. McKenna requested "one of those little chocolate things" with his double latte. The waiter's brow knotted again. Never mind, McKenna said wearily, just bring the double latte.

When I told McKenna that I wasn't sure how his timewave theory worked, he launched into an explication of it. The essence of the theory is that existence emerges from the clash of two forces, not good and evil but habit and novelty. Habit is entropic, repetitious, conservative; novelty is creative, disjunctive, progressive. "In all processes at any scale, you can see these two forces grinding against each other. You can also see that novelty is winning."

As novelty increases, so does complexity. From the big bang on, "the universe has been complexifying, and each level of complexity achieved becomes the platform for a further ascent into complexity. So fusion in early stars creates heavy elements and carbon. That becomes the basis for molecular chemistry. That becomes the basis for photobionic life. That

---

*McKenna's doubts about Ram Dass's story in *Be Here Now* (Crown Publishing, New York, 1971) seemed to be confirmed by a recent article: "The Evolution of Just Plain Ram Dass," by Paul Krassner (*High Times*, June 2001, p. 90). Krassner, a self-described old friend of Ram Dass, stated that Ram Dass "now admits that he made up the story he told American seekers about the time he gave his guru in India three tablets of LSD and nothing happened." I called Ram Dass to confirm the story, and he denied that he had ever told Paul Krassner or anyone else that the story wasn't true. When I contacted Krassner to inform him of Ram Dass's response, he said that to the best of his recollection the anthropologist Stanley Krippner had told him that Ram Dass had taken back the guru story. When I contacted Krippner, he said that Krassner must have misunderstood him; he, Krippner, had no reason to doubt Ram Dass's story.

becomes the basis for eukaryotic stuff. That becomes the basis for multicellular organisms, *that* for higher animals, *that* for culture, *that* for machine symbiosis, and on and on."

Modern science often depicts humanity as an accident, a bit player in the universe, but the timewave theory puts us at center stage in the cosmic drama. "A million years ago, Darwinian evolution was what was happening. But now it's human social systems, human technological advancement." If McKenna had to define God, he would define it as the novelty-generating process. This definition can serve as the basis for a new moral order. "Anything which destroyed novelty would be bad, and anything which helped build it up and advance it would be good." He stirred his double latte.

What about Nazi Germany? I asked. Wasn't that novel? Or the hydrogen bomb? Or AIDS? McKenna acknowledged that novelty may be accompanied by increased suffering and death. For example, "a huge epidemic which wipes out a whole ruling class creates novel situations." But in general, progress of some kind emerges from these catastrophes. In the case of Nazi Germany, "the twentieth century *had* to deal with the issue of fascism. It couldn't close its eyes and waltz past that. And it did! So in that sense Nazi Germany, with its science-fiction production values and its silly rhetoric, served a useful purpose." In other words, *things are getting better.*

As early as the 1970s, McKenna sought to make his drug-inspired insight precise and quantitative. He discovered that fractals, mathematical objects whose patterns repeat themselves at different scales, provide an excellent model of the entropy-novelty dialectic. "The fall of the dynasty, the collapse of the love affair, the ruin of the corporation, and the death of the amoeba are all somehow dramas with the same energy points and flows embedded in them."

So what does McKenna *really* think is going to happen on December 22, 2012? "If you *really* understand what I'm saying," he replied, "you would understand it can't be said. It's a prediction of an unpredictable event." The event will be "some enormously reality-rearranging thing." Scientists will invent a truly intelligent computer or a time-travel machine. Perhaps we will be visited by an alien spaceship or an asteroid. "I don't know if it's built into the laws of spacetime, or it's generated out of human inventiveness, or whether it's a mile and a half wide and arrives unexpectedly in the center of North America."

But does he actually think the apocalypse will arrive on December 22,

2012? "Well . . ." McKenna hesitated. "No." He has merely created one mathematical model of the flow and ebb of novelty in history. "It's a weak case, because history is not a mathematically defined entity." His model is "just a kind of fantasizing within a certain kind of vocabulary." He still believes in the legitimacy of his project, even if his particular model turns out to be a failure. "I'm trying to redeem history, make it make sense, show that it obeys laws."

But he couldn't stop there. His eyes wide, he divulged a "huge — quote unquote — coincidence" involving his prophecy. After he made his prediction that the apocalypse would occur on December 22, 2012, he learned that thousands of years ago Mayan astronomers had predicted the world would end on the very same day. "And now there has been new scholarship that they were tracking the galactic center and its precessional path through the ecliptic plane. What does all this mean?" McKenna leaned toward me, his eyes slitted and his teeth bared. "It means we are trapped in software written by the ghost of Jorge Luis Borges!" He threw his head back and cackled. "Tell *that* to the National Academy of Sciences!"

The Matrix and "The Zahir"

In 1996 geneticists at the National Cancer Institute announced that they had found a gene associated with novelty-seeking, or "thrill-seeking," as some journalists put it.* The researchers said that the novelty-seeking gene is highly correlated with risky behavior, including consumption of illicit drugs. The evidence for the novelty-seeking gene is rather flimsy, but if there is such a gene, McKenna obviously has it. He exemplifies what Huston Smith calls (disapprovingly) "neophilia," an obsessive "embrace of the new." McKenna is so enamored of novelty that he deifies it. Just as warm-hearted Huston Smith discerned a loving God during his psychedelic trip, McKenna perceived a novelty-generating God.

---

*See "Population and Familial Associations Between the D4 Dopamine Receptor Gene and Measures of Novelty Seeking," by J. Benjamin et al. (*Nature Genetics,* vol. 12, 1996, pp. 81–84). At least two other groups have reported finding no evidence for the novelty-seeking gene. According to James Austin (*Zen and the Brain,* MIT Press, Cambridge, 1998, p. 219), experiments in rats suggest that exposure to novel situations generates a pulse of the endogenous opioids known as endorphins. Novelty-lovers may be motivated to seek novelty because they are rewarded with larger-than-average pulses of endorphins; or, conversely, they may seek out more intense thrills because they get a smaller endorphin rush than others do from the same stimulus.

McKenna's theology, when stripped of its extraterrestrial, psychedelic, apocalyptic trappings, is surprisingly conventional. Reality, he said, "has a strange artfulness to it that betrays the hand of a kind of director, or author, or some kind of intelligence which is shaping this supposedly chaotic and random thing." This is the old Christian argument from design, which holds that reality cannot be merely the result of random acts of nature; it must have been deliberately created, to fulfill the divine plan.

The ultimate purpose of the plan, as McKenna sees it, is not the salvation of our souls but the production of novelty for our delectation. Although we have faced enormous challenges — Nazism, thermonuclear weapons, AIDS, terrorism, global warming — somehow we will prevail. In fact, these horrific events serve a divine purpose: They generate newness. Like Stanislav Grof's thicken-the-plot theodicy and Freeman Dyson's principle of maximum diversity, McKenna's timewave theory values drama, adventure, and excitement over peace, freedom from suffering, nirvana.

McKenna occasionally gives his theology a gnostic twist, just so no one can accuse him of being sentimental. He saw gnostic overtones in the proposal of certain physicists and artificial intelligence enthusiasts that our reality might actually be virtual reality, a digital simulation created by an alien and possibly malevolent intelligence. "The dawning of this metaphor of virtual reality and all that is permitting a new kind of paranoia," he said, "more exotic than any previously entertained, a paranoia that reality itself is some kind of software."

The Matrix brilliantly dramatizes this concept, McKenna said. The 1999 film depicts a future in which evil robots have conquered us and stuck us in beakers, where we serve as organic batteries providing power for the robots. The robotic demiurges keep us pacified by electronically stimulating our brains to make us believe that we still live in our old, relatively benign, pre-robot world. (The movie's biggest weakness is that robots clever enough to crack the neural code, the secret language of human brain cells, cannot invent a more efficient energy source than humans in big beakers.)

Although *The Matrix* is an expression of gnosticism at its most paranoid, it holds out the hope of enlightenment, or gnosis, a flash of supreme knowledge that liberates us from this demonic world. The movie's hero, played by Keanu Reeves, takes the first step toward gnosis when he rips the electrodes from his skull and realizes that what he believed is reality is actually "the matrix," a virtual reality created by the evil robots. At the end of the movie Reeves achieves total gnosis — and God-like power within the matrix — by fully recognizing its illusory nature.

McKenna parted company with the movie on this point. He guessed that gnosis may always be a fleeting experience that eludes capture by any language or system. "The real truth is" — he grimaced with the effort — "beyond human grasp. It's somehow inexpressible. You can confront the truth, you can know the presence of God. But you can't write about it, talk about it, and get it right. It somehow is translinguistic, without being unitary, or some kind of Neoplatonic buzz or something. No, it's very complicated. But it somehow fills the interstices of language."

McKenna's love of novelty, adventure, the quest, I suspected, made him ambivalent toward gnosis. No one expressed this ambivalence toward ultimate knowledge better than Borges. In his story "The Zahir," the narrator is a man who becomes increasingly obsessed with a coin that he receives from a shopkeeper. Although he attempts to rid himself of it, the coin keeps coming back into his possession. Eventually he cannot think about anything but the coin; it even haunts his dreams. As the story unfolds, he gradually realizes that the coin is what ancient Islamic texts called a Zahir, an emblem of the mystery of existence, the riddle at the heart of things.

At the end of the story, the narrator is still sane enough to foresee his fate: "I shall pass from thousands of apparitions to one alone: from a very complex dream to a very simple dream. Others will dream that I am mad, and I shall dream of the Zahir. And when everyone dreams of the Zahir day and night, which will be a dream and which the reality, the earth or the Zahir?" In Borges's eerie fable, gnosis — a vision of the one thing that explains all other things — represents not salvation or omnipotence or another crest of the timewave, but a black hole into which we vanish forever.

Facing the final trip

Two weeks after I met him in New York, just after he returned to his home in Hawaii, McKenna collapsed in the throes of a seizure. His girlfriend, Christy Silness, managed to get him to a hospital in Honolulu, where tests revealed an enormous malignant tumor in his brain. McKenna's choices were grim. The physician recommended gamma-ray surgery, in which converging beams of radiation bombard the tumor. This treatment might give McKenna another six months to a year, but it could also cause dementia and other side effects. Untreated, he would probably die within a month. McKenna chose the radiation surgery.

His fans tried to restore his health through nonmedical interventions. In one attempt, Art Bell, a popular radio talk-show host, exhorted his thirteen million listeners to pray for McKenna's good health. McKenna chose not to pursue other unconventional remedies requiring his personal involvement. "There are various options when you are faced with a terminal disease," he told a reporter for *Wired*. "One is cure-chasing, where you head off to Shanghai or Brazil or the Dominican Republic to be with these great maestros who can save you. The other thing is to do what you always wanted to do. So that means head to Cape Canaveral to see a shuttle launch, on to sunrise over the pyramids, on to a month in the Grand Hotel de Paris. I wasn't too keen on that either. My tendency was just to twist another bomber and think about it all."

During my lunch with McKenna at the Millennium Hotel, I asked him if all his psychedelic excursions had mitigated his fear of death. His reply revealed how hardheaded he was beneath all the phantasmagoric blarney. "I wouldn't say I have no fear of death," he said. "I am *interested*. Ultimately, my assumption is that, if I have the opportunity, I would embrace it if I saw it coming. But I'm scientific in my approach to my own knowledge of death. In other words, DMT may show you what the dying brain is like. It could be like some kind of necreptogen or something like that. But dying is not death. Near-death experiences are not *death* experiences."

McKenna responded well to the gamma-ray treatment. He made it past the millennial cusp, but he went downhill rapidly after that. He died on April 3, 2000, less than eleven months after I met him. He was fifty-three. Just before his death, another psychedelic advocate told *Wired* that McKenna's outrageousness, like that of Timothy Leary, may have harmed the psychedelic cause: "Some people would certainly argue that it doesn't help to have the most famous second-generation psychedelicist be another man in a purple sparkly suit."

McKenna's attempts to serve as a serious advocate for psychedelics were no doubt undermined by his irony and wit and Loki-like mischievousness, his penchant for Borgesian fantasy, but those were precisely the qualities that I found so appealing in him. My reaction to McKenna was the opposite of my reaction to Ken Wilber. I objected to much of what

McKenna said — at least if interpreted literally — but I loved the way he said it. Wilber pays lip service to wonder, but to McKenna wonder was the essence of gnosis. As he told me during our interview, all his confabulations were intended to make us see that the world is "a weird, weird place."

McKenna feared, perhaps with good reason, that we have all become as obtuse as the waiter who served us lunch atop the Millennium Hotel. To shake us from our perceptual torpor, McKenna played the holy fool, the crazy-wisdom adept. He pushed our faces in the exotic, lurid inventions of modern science and technology, including superstring theory, time travel, virtual reality, artificial intelligence. He even stooped to speculating about extraterrestrials and to forecasting the end of life as we know it. What elevated him above most other prophets was that he delivered his prophecies with a wink, an implicit acknowledgment that ultimately reality is stranger than we can say or imagine.

The Nature-Does-Not-Care Principle

In *The Gnostic Religion,* first published almost half a century ago and still a definitive treatment of the subject, the philosopher Hans Jonas argued that gnosticism foreshadowed the alienated outlook of existential philosophers such as Friedrich Nietzsche and Jean-Paul Sartre.* But as dark as gnosticism was, Jonas pointed out, existentialism is much darker. Gnosticism postulated that the world was created by an insane or evil Demiurge, but even a flawed God is better than no God at all. "In the gnostic conception the hostile, the demonic, is still anthropomorphic, familiar even in its foreignness," Jonas wrote. Modern existentialism, which depicts the cosmos not as hostile but as utterly indifferent to humanity, is "infinitely more radical and more desperate than gnostic nihilism could ever be. That nature does not care, one way or the other, is the true abyss."

* Sartre embraced atheism at the age of ten, but his bleak view of existence may have been exacerbated when he was thirty by a hellish drug trip. According to "Two Classic Trips," by Thomas Riedlinger (*Gnosis,* Winter 1993, pp. 34–41), Sartre persuaded Daniel LaGache, a friend and physician, to inject him with mescaline in 1935. During the trip Sartre was attacked by giant octopuses, apes, and insects. For almost a year afterward, he suffered flashbacks in which monstrous lobsters chased him, and he feared he was losing his mind. The story had a happy ending: According to Riedlinger, Sartre's trip inspired his nightmarish novel *Nausea,* which was published in 1938 and helped him to win the Nobel Prize in literature twenty-six years later.

One eloquent spokesman for the nature-does-not-care principle is Steven Weinberg, a Nobel laureate and one of the twentieth century's most prominent physicists. Weinberg scoffs at the claim — advanced not only by provocateurs like McKenna but also by some respectable physicists — that modern physics is revealing a "plan" underlying reality, a plan that gives us a central role. Some of Weinberg's colleagues, intoxicated by their ability to discern profound symmetries in nature — symmetries embodied in quantum field theory and the theory of general relativity — have proposed that "God is a geometer." By uncovering nature's geometric structure, these physicists suggest, they are unveiling "the mind of God." Weinberg retorts, in effect, that if God is a geometer, He is a geometer who couldn't care less about us.

Weinberg's main objection to belief in God is the problem of natural evil. His own life has been "remarkably happy, perhaps in the upper 99.99 percentile of human happiness," he emphasized in a recent essay. But given the suffering he had seen meted out to others, Weinberg did not understand how any rational person could believe in a God who cares for us. "I have seen a mother die painfully of cancer, a father's personality destroyed by Alzheimer's disease, and scores of second and third cousins murdered by the Holocaust." He rejected the proposition of some theologians that evil is the price we pay for God's having given us free will. "It seems a bit unfair for my relatives to be murdered in order to provide an opportunity for free will for the Germans, but even putting that aside, how does free will account for cancer? Is it an opportunity of free will for tumors?"

I bet that Terence McKenna, if pressed, would have agreed with Weinberg that the nature-does-not-care principle is quite plausible. As McKenna once said of his timewave theory:

> The notion of some kind of fantastically complicated visionary revelation that happens to put one at the very center of the action is a symptom of mental illness. This theory does that, and yet so does immediate experience, and so do the ontologies of Judaism, Islam, and Christianity. My theory may be clinically pathological, but unlike these religious systems, I have enough humor to realize this.

Here McKenna, it seems to me, was alluding to the scandal of particularity. Any theology putting us at the "center of the action" is "pathological," a symptom of our compulsive narcissism and anthropomorphism.

Of course, all theologies put humanity at the center of the action, including not just the Western faiths and McKenna's timewave theory but also Hinduism, Buddhism, the perennial philosophy, Grof's thicken-the-plot theodicy, gnosticism, and my own humble variant thereof. The alternative, as Steven Weinberg has proposed, is that there is no plan, or at least not one in which we play a role. Nature does not care about us. McKenna's eschatological obsessions, his prophecies about the "final time" and "the end of the wave," might have been prompted by his recognition of this bleak possibility: We appeared through sheer happenstance, and we could vanish in the same way.

Not long after McKenna died, I read another obituary in the *New York Times* that made this point quite brutally. The obituary was for Jeffrey Willick, an astrophysicist at Stanford. He was sitting in a café, sipping coffee and plinking away at a laptop, when an out-of-control car crashed through the front window and killed him. Willick was forty, married, with two children, six and four years old. His wife was expecting another child. It was Father's Day.

One of Willick's professional interests was whether there is an element of randomness in the motion of galaxies relative to each other. His concern with cosmic caprice was excruciatingly appropriate. We would all like to believe that God or fate or whatever would not allow us or those we love to die in such a senseless, untimely manner. Yet as the life and death of Willick demonstrated, the end can come at any time, in the form not only of a car crashing through a window or a brain tumor, but also of an asteroid or supernova shock wave. Such an event would not thicken the plot. It would end the plot.

11

Ayahuasca

There are moments when I teeter on the edge of belief that nature cares. The occasion may be mundane. I may be raking leaves on a gray fall day, drinking a glass of wine with my wife, Suzie, on our deck at sunset, waiting with my son and daughter at the end of the driveway for the morning school bus to arrive. Gratitude wells up in me as a kind of yearning, as strong as hunger or sexual desire. I want to thank someone, something, for all that I have. This sensation was particularly powerful during the magical summer that my family spent with Lena, the crow foundling raised by Suzie.

A God who deserves thanks for my good fortune, I had to remind myself, also deserves blame for the misery of countless others. Thanking this God for all I have would be obscene. I would be saying, in effect, "Thank you, God, for not screwing me like you've screwed all those other poor bastards." I would be no better than the devout warlord who thanks God for helping him slaughter the infidels. A God who dispenses favors so capriciously must be profoundly flawed, if not demonic. He certainly does not deserve our gratitude or adulation. I'm not blessed, I'm just lucky, very, very lucky, and my luck may turn at any moment. Lena's death seemed to confirm life's accidental nature.

On the day after Lena died, I had to travel to California to take ayahuasca. I first heard about the Amazonian hallucinogen called yage back in my youth. Only when I began my research for this book did I discover that yage, now more commonly called ayahuasca, was becoming the entheogen of choice for many spiritual seekers. The more I learned about ayahuasca, the more it intrigued me for both personal and journalistic reasons. When I made contact with a group offering guided sessions, I eagerly signed up.

But as a plane bore me westward, I dwelled on the risks that ayahuasca posed, physical as well as psychological.* I also felt guilty for abandoning my grief-stricken family, and particularly Suzie. I had not seen her so distraught since one of her dearest friends died of AIDS several years earlier. In part to assuage my guilt, before I left home I promised her that during my ayahuasca session I would find some sort of sign or meaning or *something* that would make her feel better. I was afraid it was a promise I couldn't keep.

Given my mood, I felt fortunate that I would not be taking ayahuasca for two days, and that I had arranged a meeting after my arrival in California with Ann and Alexander Shulgin. This would be a propitious way to begin my journey, because between them the Shulgins probably knew as much about psychedelics as any couple alive. If anyone could give me guidance and reassurance about my upcoming encounter with the vine of the dead, they could.

Alexander "Sasha" Shulgin was a top-rank researcher for Dow Chemical in 1960 when he ingested a psychedelic compound, mescaline, for the first time. Shulgin found the experience so astonishing that he devoted the rest of his career to psychedelic chemistry. He left Dow in 1966 and supported himself thereafter by consulting, lecturing, and teaching. Working out of a laboratory on his ranch east of San Francisco, he synthesized more than two hundred novel psychotropic compounds. Although he did not discover methylenedioxymethylamphetamine, also known as MDMA or Ecstasy, Shulgin is credited with — or blamed for — popularizing it by drawing attention to its empathogenic properties in the 1970s.

---

*Ayahuasca "is not even remotely a recreational drug. It is an extremely potent hallucinogen which no one should ingest carelessly, or without a full understanding of how it works in the body," the psychedelic botanist Jim DeKorne warned in *Psychedelic Shamanism* (Loompanics Unlimited, Port Townsend, Washington, 1994, p. 93). Ayahuasca contains monoamine oxidase inhibitors, which can cause severe headaches, cardiac arrhythmia, and pulmonary edema when combined with the proteins in such common foods as cheese, beer, wine, dried fruit, yogurt, soy sauce, coffee, chocolate, and pickles, according to *Psychedelic Shamanism*. Nor should ayahuasca be combined with any amphetamine-type compounds, including ephedrine, epinephrine, or MDMA. Mixing ayahuasca with so-called selective serotonin reuptake inhibitors, the best known of which is Prozac, can cause dangerous and possibly fatal reactions, according to "Ayahuasca Preparations and Serotonin Reuptake Inhibitors: A Potential Combination for Severe Adverse Reactions," by James Callaway and Charles Grob (*Journal of Psychoactive Drugs*, vol. 30, no. 4, pp. 367–369). The surest way to avoid these physiological side effects is to fast for twelve or more hours before consuming ayahuasca, but even so, it is likely to make you nauseated.

After synthesizing the compounds in his laboratory, Shulgin tested them on himself, Ann, and a group of trusted friends. These "psychonauts" took meticulous notes on their research sessions and rated their experiences on a scale invented by Shulgin. It ranged from a minus sign, which represents no change, up to plus four (written as ++++), which is a sublime, potentially life-changing, "peak" experience.

There were a few rules for the sessions. Subjects could not be taking any medication, and they had to refrain from ingesting any other drugs for at least three days before a session. If someone said "Hand in the air" while raising her hand during a trip, that meant she wanted to discuss a serious "reality-based concern or problem" (for example, a smoky smell in the kitchen). Sexual contact was prohibited between people not previously involved. "Of course, if an established couple wishes to retire to a private room to make love, they are free to do so with the blessings (and probably the envy) of the rest of us," Shulgin once remarked.

Although Shulgin wrote technical papers on MDMA and other drugs for peer-reviewed journals, the bulk of his findings were confined to his personal notes. In the early 1990s Sasha and Ann poured their knowledge into two fictionalized memoirs, called *PIHKAL* and *TIHKAL*, published under their own imprint. The titles are acronyms for "phenethylamines I have known and loved" and "tryptamines I have known and loved." Phenethylamines and tryptamines are the two major classes of psychedelic compounds. Together the two books tell the tale of a decades-long love affair of two highly intelligent, bohemian protagonists: "Shura," a chemist, and "Alice," a writer and lay psychotherapist. They share a passion not only for each other but also for mind-expanding drugs. In alternating chapters, Shura and Alice reveal how their lives have been intellectually, sensually, and spiritually enriched by their pharmacological adventures.

Both *PIHKAL* and *TIHKAL* also present detailed recipes for scores of phenethylamines and tryptamines and descriptions of their physiological and psychological effects. "No one who is lacking legal authorization should attempt the synthesis of any of the compounds described in the second half of this book," the Shulgins warned in a "Note to the Reader." They nonetheless declared that investigations of the scientific and therapeutic potential of psychedelics "must be not only allowed but encouraged. It is essential that our present negative propaganda regarding psychedelic drugs be replaced with honesty and truthfulness about their effects, both good and bad."

In chapters titled "DMT Is Everywhere" and "Hoasca vs. Ayahuasca," *TIHKAL* offered a wealth of information on DMT, ayahuasca, and other DMT-containing plants. To illustrate "the agony and ecstasy of ayahuasca," the Shulgins presented a first-person account by an anonymous psychonaut who had ingested a homemade ayahuasca-type brew. Within a half hour, the narrator was overwhelmed by fantastical geometric hallucinations, along with nausea and an urge to defecate. Stumbling to a bathroom, he sat down on a toilet. Closing his eyes, he found himself rocketing down "incredibly complex and insane roads" that he could "never fully describe." When he opened his eyes and looked down, he found that his legs had disappeared. "Oh no," he thought, "it's OK. There they are. No, they're gone again." Reading this account, I had to wonder: Why do we do this to ourselves?

Advice from Ann and Sasha

The directions that Sasha had faxed me were detailed and meticulous, like his recipes for synthesizing hallucinogens. After driving for about an hour into the foothills east of San Francisco, I made my way down a dusty dirt road to a tree-shaded, one-story house with several outlying sheds, one of which served as Sasha's laboratory. Sasha was a big, barrel-chested, rugged man with a hoary, leonine beard and mane. Ann's lined face and crinkled eyes gave her a tender, empathetic expression.

We sat in a living room crammed with books, papers, and potted plants. Pinned to one wall was a piece of yellow tape that read "SHERIFF'S LINE: DO NOT CROSS." It was a memento of a 1998 raid by the local sheriff's department, which had wrongly suspected Sasha of manufacturing methamphetamine, also known as "crystal" or "ice." Shulgin's work has always been legal; the Drug Enforcement Administration has licensed him to do research on scheduled compounds. But now and then, he said, overzealous or ignorant law-enforcement officials harass him anyway.

A pattern emerged early on in my conversation with Ann and Sasha. I asked, Do you think the legal and political climate for psychedelics is improving? No, Sasha replied, shaking his head. If anything, things were getting worse. He was appalled by a recent federal law giving the police power to confiscate the property of those accused of breaking drug laws.

"I have a different view on that," Ann said. She was encouraged by the fact that commentators, or at least intelligent ones, increasingly refer to the "failed" war on drugs. "Everyone knows this thing has not only failed, it has made the drug problem actually worse," she said. "If we get one politician with courage, that's all it's going to take to break the whole thing apart and start changing things."

"She's optimistic, I'm pessimistic," Sasha said. "We balance out very nicely."

Later, Ann said she firmly believed in reincarnation. Sasha found reports about people remembering their past lives interesting but ultimately unconvincing. Ann intuited a divine intelligence guiding the cosmos, while Sasha was skeptical. She was the romantic empathizer, he the rationalist. She was the psychotherapist, he the chemist. But they were unfailingly gracious toward each other. When Ann interrupted Sasha to disagree with him, as she often did, he seemed less irritated than charmed.

Sasha liked to turn my questions back on me. What did I mean by mysticism? By God? When I asked if he meditated, he replied that it depended on my definition of meditation. "Are you *doing* things with your mind, or are you *undoing* things? Structuring, or destructuring? Assembling and analyzing, or disassembling and avoiding?"

Sasha had tried Zen but found no benefit in it. "The idea of sitting there quietly and voiding your mind of any thoughts, of any process, of turning off the record, just turning the amplifier not down but *off*— I find it frightening! I don't see what the virtue is. You're in absolute, thoughtless, mindless space for about twenty seconds. And I say to myself, Why the hell am I doing this?" If meditation means total immersion in an activity, being absorbed in the moment, Sasha continued, well, he does that whenever he works in his laboratory. "I consider that meditation, but *very* active," he says. "For me that's a treasure."

When I asked Sasha how many drug trips he had taken in all, he said it depended on how I defined *trip*. When exploring a new compound, he would start with very small amounts, to test for potency, and gradually increase the dose. "Not all of these were trips, and a lot of them were just exploring." He had taken compounds that were at least potentially psychoactive three or four times a week for more than forty years, but only a few thousand of those experiments were genuine trips.

Their psychedelic days were over, Sasha and Ann assured me. Ann had once given MDMA to her psychotherapeutic patients, but she stopped after the drug was outlawed in 1986, under the so-called Designer Drug Act. The team of psychonauts who had tested compounds concocted by Sasha had disbanded. Sasha said his research continues; one of his current projects involves searching for new antidepressants. But he no longer ingests or synthesizes psychedelics.

Like other spiritual practices, psychedelics are a two-edged sword, Sasha said. They may help us become more compassionate and wise, but they may also lead to ego inflation or worse. He posed a hypothetical question: What if a psychedelic drug helps an evil person accept his evil nature? Would that be a positive step? "It's not a panacea."

I asked if either of them believes in God. Define *God,* Sasha demanded. I mumbled something about a creative force or intelligence underlying the design of the universe.

"I believe the concept of God is absolutely unnecessary," Sasha declared.

"Unnecessary?" Ann said, staring at him.

"That's a straight answer," Sasha growled. "Things are what they are."

Ann pressed him. "Do you think the concept of a purposeful universe is nonsense?"

"It's nonsense, yeah," Sasha replied. "I don't think it's created by a divine force with a beard."

No one of any intelligence, Ann told her husband sternly, takes that old patriarchal image of God seriously anymore. Turning back to me, she said she believed that some sort of intelligence or consciousness underlies material reality, but it isn't distinct from us. "We're all parts of it, expressions of it. So we *are* it."

Ann had a friend who experienced God as pure love. "That brings out the cynicism even in noncynics," she said. How can anyone believe that God is love, given how suffused nature is with pain and suffering? The answer, Ann suggested, echoing what Huston Smith had once told me, is that our suffering is somehow a necessary part of our development and learning. "It's a little bit like watching your one-year-old experimenting." When children fall down and cry, "you sympathize, because they are having a little bit of pain on their bottom. But you realize that that is a step toward growing up." Psychedelics can help you see things from this cosmic perspective.

Sasha and Ann both rejected the notion of enlightenment as a final state of mystical knowledge. There is no final state, Sasha said, only a never-ending process. Ann agreed. She had had a few flashes of what Zen Buddhists call satori, both in psychedelic visions and in lucid dreams. "But they are not a destination. They are a reminder."

I said that psychedelics can draw us in opposite directions: They can make us feel blissfully connected to all things, or alienated and alone. Which experience is truer?

"The place I think the Buddhists try and get you to," Ann responded, "is right on the knife edge between the two. That's where the truth is. But don't ever forget that the truth of the universe changes second by second. It's not the same universe it was when we sat down at this table." Our development, our learning, never stops. "You learn in your sleep, from conversations. You learn unconsciously, consciously. You learn from every book you read and every trip you take. You're experiencing and taking in and changing as a result all the time, and yet you remain the same, essentially."

Sasha gave me advice that had helped get him "through many years, and will get me through a few more": Never lose your sense of humor or take yourself too seriously.

"The laughing Buddha is your best guide," Ann added. "What the heck is he laughing about? You can't explain that logically, but you can get into that state. And the final answer you're looking for is the knife edge, because both exist: that terrible darkness and that absolute life."

I asked whether psychedelics had helped them come to terms with their mortality. Sasha said his view of death kept evolving. As a young man, he believed that when you die, you die; your consciousness is extinguished. In middle age, his fear of death became so acute that it intruded on his research. Now, at seventy-four, he looked at death as "another transition, another state of consciousness. Admittedly it's one I've not explored, but then again, any new drug is one you've not explored." He did not look forward to death, but he no longer feared it, either. Sasha spoke quietly, calmly.

But did that mean he no longer viewed death as a mystery?

"No way! It's gotten a hell of a lot more complex than I ever dreamed it would be."

Ann's psychedelic experiences had bolstered her faith that "the mind, consciousness, almost certainly exists outside of the body" and will survive death. After her brother died unexpectedly of a heart attack a year ago, Ann was overcome by grief. But when she viewed her brother's body before he was buried, her grief gave way to a strange joy, as she felt her brother's intelligent, humorous presence surrounding her.

Ann had much she wanted to accomplish before she died, but otherwise she did not fear death. "I've never believed there was nothing on the other side," she said. "It doesn't make any sense. We are continuing streams of energy. Now, the form you take afterward, the form of the consciousness, that's open to some question. But I have a feeling that we all know, because we all have the unconscious memory of having gone through it many times before. I think it is really a going home. I think it will be familiar as soon as you get to the door."

When I said goodbye to the Shulgins, I was in a much better mood than when the day began. I couldn't wait to call Suzie and tell her about this wonderful couple. It wasn't so much what the Shulgins said that impressed me — their thoughts on life and death and enlightenment and spirituality echoed what others had told me — but rather how they said it. Through the give-and-take between them, the Shulgins seemed to embody more wisdom than each individually. I was particularly eager to tell Suzie about Ann's view of death, which I found comforting; maybe it would ease Suzie's grief over Lena's death the day before.

When Suzie answered the phone, her voice was hoarse and hushed. Perhaps too eagerly, I described my afternoon with the Shulgins. I recalled what Ann had said about her brother's death, about how she saw death as not an ending but a transformation that we have undergone countless times. Coming out of my mouth, though, it sounded terribly trite.

Of course no one really has any answers about death, I told Suzie. It's a mystery, an unsolvable mystery, but Ann was so warm and sincere and empathetic that she made me feel better. Maybe this is the best we can do, I said, just to share our fears and sorrows with each other, to recognize that we're all in this together. That can help, can't it?

Suzie murmured vague assent. She had spent the day with a friend, also a bird lover. The friend had said something that made Suzie feel a little better about Lena's death: Some creatures appear among us only briefly, but in that instant they shine as brightly as shooting stars. Lena, Suzie whispered to me from the other side of the continent, was a shooting star. I thought, but did not say, We're all shooting stars.

SALUD!

Two days later, the sun was descending toward the Pacific when I turned off the coastal highway and headed inland toward my ayahuasca rendezvous. My stomach was growling; on the advice of the session leaders, I had not eaten all day. The route led east through a redwood glade, swung back west, and broke out onto a brown, treeless headland crisscrossed with wire fences. After passing a couple of dilapidated barns, I arrived at my destination, a fenced-off ranch perched on a gentle slope high above the Pacific. As I pulled up next to a half-dozen other cars in the driveway, I thought: Too late to turn back now.

Six of the nine people with whom I would spend this evening were already there. (I have changed their names to preserve their privacy.) The two owners of the ranch were Allen, founder of a health food company, who was genial, fiftyish, with receding hair, and Deborah, Allen's wife and business partner, who had short blond hair and wide-set gray eyes. Although she welcomed me warmly, Deborah seemed faintly melancholy, as if distracted by some private grief.

The other four people present were willowy, blond Linda, who rented a guesthouse from Allen and Deborah; Nancy, Linda's roommate, a firefighter with the build of a serious weightlifter; Michael, thirtyish, also an executive in the health food business, with an Irish boxer's face; and George, who had the sun-bleached hair and George Hamilton tan of an aging surfer.

About a half hour after I arrived, the final three members of our group showed up: Tony, who had dark hair and soulful, protuberant eyes; bearish, balding Kevin; and Blaed, Kevin's twenty-something nephew, who had sharp, angular features and a goatee. Tony and Kevin, both scientists with extensive knowledge of psychedelics, were coleaders of this session. They opened the trunk of their car and unloaded plastic cups, a tape player, and sleeping bags; bags of food for breakfast the next morning; and a cooler containing several large plastic bottles filled with what looked like purple-brown spit: ayahuasca.

The sun set, leaving the sky stained with bloody, Rothko-esque swaths. As night fell, the tension grew. This would be the first ayahuasca trip not only for me but also for Deborah, Nancy, George, and Blaed. At nine o'clock we headed out of the house. The sky was clear, emblazoned with stars and a nearly full moon. A hundred yards or so from the house, Nancy and Linda had created a "sacred circle," a patch of gravel ringed by fist-sized stones, where we would spend the night. At the center of the circle was an altar, a box covered with a multicolored cloth.

We took our places around the circle, setting down blankets, sleeping bags, pillows. Tony told us to put our "sacred objects" — the items of personal significance that we were supposed to bring to the session — on the altar. I took Lena's glossy black feather out of my pocket and put it on the altar beside a vase of flowers, an owl feather, an amethyst crystal, bongo drums, a leather rattle, and a tiny bust of Queen Nefertiti. Tony lit a bundle of sage and wafted the smoke around the site. "It's for purification," he said with an embarrassed grin.

Tony gave each of us a shiny new steel bucket. It would be best, he advised, to go to a nearby embankment to vomit, but if we couldn't make it that far, we should use the bucket. We would probably all get sick, but that's okay. Vomiting has a therapeutic, purgative effect. This ayahuasca, which Tony obtained from a Brazilian sect, was the best he had ever sampled. It had six times more DMT than the average batch, according to a chemical analysis done by a friend of Tony's.

Members of the Brazilian church usually take 50 milliliters, about a quarter of a cup. They need only a modest dose, Tony explained, because ayahuasca consumption leads to reverse tolerance; over time, smaller amounts produce the same psychotropic effects. Tony was giving us 120 milliliters, because he wanted to be sure we would have a strong, satisfying experience. Better to have too much ayahuasca, he said, than not enough. With smaller doses, you might merely get sick without experiencing any hallucinations.

Tony held up a rope woven of multicolored strands, which he said symbolized the unity of our group. He asked each of us to tie a knot in the rope and say something. Tomorrow morning we would untie our knots and share our thoughts again. Taking the rope, Michael said he wanted to find ways to get closer to his family, including a brother from whom he was estranged. Allen hoped to come to terms with the recent death of his father. Kevin and his nephew Blaed were concerned about a relative who had cancer. I said that my wife was very upset because an animal she loved had just died; I hoped this session might teach me something that could console her. Linda said that she was not taking ayahuasca tonight; her goal was to help the rest of us in any way she could. Grasping a bottle, Tony poured some of the contents into a steel measuring cup and then into nine green plastic cups. There was plenty of ayahuasca to spare, he assured us. Anyone who felt no effects after an hour or so should ask for a 50-milliliter supplement then, or at any time during the night. He suggested that we all drink at the same time. Following his lead, we stood and faced one another, everyone except Linda holding a cup. "Salud!" Tony exclaimed, and as one we quaffed our doses.

Meeting the DMT Demiurge

It tasted astonishingly foul, an unwholesome mixture of bitterness and rancid sweetness. I gagged, as did almost everyone else. Tony urged us to suppress our nausea for as long as possible, to give the ayahuasca time to take effect. Blaed said it tasted like stale dregs of stout. Dregs with cigarette butts in them, someone added.

The conversation died down. Some of the group remained standing, others sat or lay on their sleeping bags. I stood looking east at the hills, the stars, the moon. The Pleiades were diamond chips flung across a strip of gauze. A planet — Jupiter, someone said — blazed in the southern sky. A plane or a satellite hurtled overhead. A star caught my eye, hovering about twenty degrees above the horizon. It twinkled at the end of a long tube or tunnel, beaming across light-years of vacuum here to me. Energy radiated from the star in a shifting snowflake pattern. Was this strange, or was I just paying closer attention than usual to an ordinary low-light visual effect?

Tony turned on a tape of a Peruvian rain forest. The yelps, hoots, howls, ululations, clicks, and rattles mingled with local noises — crickets chirping, a dog barking, humans sighing, clearing their throats, grinding gravel under shoes or rear ends. The sounds had a muffled, reverberatory quality, as if we were all sealed inside a cavernous metal tank. I felt weak-kneed, dizzy, probably just from hunger, I thought. I sat and closed my eyes.

At the bottom of a dark, radiant well shimmered a vaguely heart-shaped manifold slowly rotating. Definitely strange, I thought, not your ordinary rods-and-cones hallucination.* The manifold dissolved into an incalculably more complex scene, an iridescent, alien landscape reminiscent — some pedantic part of my brain remarked — of the paintings of the French surrealist Yves Tanguy. The landscape was populated with bizarre geometric

---

*In *True Hallucinations* (HarperSanFrancisco, 1993, p. 128), Terence McKenna said that while under the influence of psilocybin mushrooms he once had a vision of a "beautiful, recursive geometric form" that "looked somewhat like a valentine or a bleeding heart." The form reminded the intoxicated McKenna of the second-century gnostic Valentinius, who held that the world was accidentally created by a female Demiurge named Sophia.

objects — shards, scimitars, French curves, manifolds — simultaneously two- and three-dimensional and lacquered, glazed, suffused with achingly lovely colors.

The jungle hoots and screams and hisses and rattles grew insistent, manic, urgent, and they seemed to inject energy into my visions, making them percolate and convect at an ever more furious rate. The forms shifted, tumbled, quivered, danced with a kind of mischievous intelligence. They were showing off, trying to stagger me with increasingly ostentatious displays of otherworldly beauty. Look at *this!* Okay, now check *this* out! But that doesn't compare to . . . *this!*

Overwhelmed, I opened my eyes. There were Deborah and Linda sitting across from me, bundled up in blankets. Silhouetted against the sky, they looked ancient, archetypal, like stoic Aztec women. There, to my immense relief, were the sky, the stars, the moon — altered, to be sure, phosphorescent plankton adrift in an opalescent sea, dewdrops in a cosmic spider's web. But they were there, they were real, and I felt grounded, back in touch with the world of things.

But then this world, too, grew strange. Flashes on the horizon, followed by ominous booms. What the hell was that? Thunder? Artillery fire? The beginning of the end? Real or hallucination? It was too much. I closed my eyes and the iridescent polygons rushed back at me with a vengeance, mutating into still more impossibly beautiful forms, as if to say, Where have you been? How dare you leave us! Behold our power!

I remembered Suzie, my vow to find something in the ayahuasca underworld that would bring solace. But whatever was putting on this display for me — this mischievous DMT Demiurge — brushed my pathetic human concerns aside. Your loves and fears are irrelevant here, it seemed I was being told. Forget them! Look at *this!* Three-dimensional, four-dimensional, infinite-dimensional manifolds in paradisiacal hues.

Waves of vertigo washed over me. I was hurtling backward through space with all these clattering, jabbering shards careening after me. To escape them, I opened my eyes again, but there was no escape, the sky was in turmoil too, beaten to a froth by the throbbing rain-forest cacophony. I heard someone retching far, far away, and I thought: At least I'm not sick.

My stomach convulsed and surged up my gullet. I grabbed my bucket just in time to catch a bolus of vomit, which flopped into the bucket like a jellyfish. I lurched to my feet, stumbled to an embankment near the circle, fell to my knees, and retched again as my head exploded into polychromatic streamers. All around me was a riot of color; the world dissolved into undifferentiated Day-Glo goo. A ten-foot pine tree at the bottom of the embankment quivered like a flame, fierce, fractal, exultant, discharging an unholy blue light. My head, too, sparked and crackled with electricity. Too much, I thought. I'm losing my mind. Too much. With a pang of guilt and horror, I thought, What if I go mad and lose my family?

Some spidery thing softly dropped onto my shoulder. Turning, expecting the worst, I saw only a human hand, a human face peering at me with concern. It was Tony. Are you all right? he asked. Yes, I said, and to my surprise I realized that I *was* all right. Some imperturbable part of me stood in the eye of the storm, calmly observing the chaos whirling around it.

Are you feeling the effects yet? Tony asked. I stared at him and emitted a grunt of incredulity. I'm blown away, I managed to say. Aren't you? Not yet, he said, shaking his head. I staggered back to my spot in the circle of stones. I felt purged, as Tony had promised earlier, but weak, rubber-kneed. I crawled into my sleeping bag and lay on my back.

With a crunch of gravel, Tony sat beside me. A few others were sitting or lying around the circle. The rest had wandered away, mostly to retch, as far as I could tell from the sound effects. Out in the darkness someone, a woman, was alternately laughing and moaning orgasmically: *Mmmm. Mmmm!* The music changed. The jungle sounds gave way to a flute ululating amid the tintinnabulation of a million microscopic cymbals. The flute was mournful, plaintive, frail, lonely; it was humanity, lost, wandering through the oblivious, crystalline cosmos.

Tony shook a rattle and sang in an alien language. His voice was sweet, pure, poignant, incomprehensible. Sitting up, I tried to hum along with him, but the noises coming from me sounded strangled, grotesque. I lay back and listened. Affection for Tony, for myself, for all of us welled up in me. We were all so tragic, comic, noble, brave, pathetic, blessed, doomed.

Awash in these sentiments, again I closed my eyes and tried to carry out my mission. A dark, winged shape hovered briefly above me, blacking out the stars. A crow spirit? Lena? The angel of death? A cloud? I tried to envision Suzie's face, her body, but she kept dissolving into polychromatic polyhedra, as if some demented art school instructor were slicing up her body into its constituent forms.

A voice: Tony, asking if I wanted more ayahuasca. Only an hour had passed! No, thank you, I said. Tony approached others around the circle and asked the same question. Only Blaed, the youngest of us, accepted the supplement. My nausea had vanished, and with it my previous anxiety. I felt devoid of emotion, as if my frontal cortex were decoupled from my emotion-generating limbic system. Although the hallucinations deepened in intensity, I watched them now with detachment, even making dry intellectual observations. The visions were not organic, animalistic, jungly, as I'd expected. They were cartoonish, mechanistic, futuristic, science-fictionish. I recalled Terence McKenna's descriptions of the entities he encountered on DMT trips: "merry elfin, self-transforming, machine creatures," "friendly fractal entities," "self-dribbling Fabergé eggs on the rebound."

But there were no alien creatures *in* the landscape before me; the whole landscape was alien. And there were no forms in space; form and space were inextricably tangled, and awash with unutterable, tip-of-the-tongue meanings. Another McKennaism came to mind, that through a strange synesthesia DMT renders syntax visible, the logic underpinning language and thought.

The colors became ever more dazzling, the shapes ever more complex, until there were no shapes and colors anymore. They yielded to something deeper and more fundamental than shape, color, syntax, thought: the metaphysical principles underlying all things, the machine code of reality. It occurred to me — or rather, to the cool, unemotional, pure intellect that I had become — that the particle physicists are right after all: God is a geometer, an infinitely intelligent, infinitely creative, utterly inhuman geometer.

The End of Life as We Know It

The soundtrack shifted again, to a bass droning interrupted sporadically by klaxons and what sounded like metal sheets being shaken and scratched. It was a demented dirge, a soundtrack for a psychotic disintegration, a descent into the underworld. It was annoying.

Desperate to get away, I crawled out of my sleeping bag and stood, wobbly-legged. Only two other people were in the circle, encased in their sleeping bags; everyone else had scattered. I had a sudden urge to see the moonlit Pacific. I headed away from the circle, down the driveway to the ranch's front gate, a hinged contraption of metal bars wrapped in chicken wire. The gate was closed. It was the only exit from the property, which was surrounded by a six-foot fence topped by barbed wire. I pushed the gate, rattled it. I poked and prodded a boxlike latch keeping the gate locked — in vain. I decided to climb the gate, then hesitated. It looked so spidery, delicate. What if I damaged it?

I could see a few figures a couple of hundred yards away, silhouetted against the sky, so I knew escape was possible. But how? My situation was assuming metaphysical dimensions when Linda materialized beside me like a guardian angel. Linda, who had abstained from taking ayahuasca, smiled as I explained my dilemma. She flipped open the infernal latch, swung the gate open for me, and bid me farewell.

Free at last, striding happily down the rutted road, I soon approached Tony, Kevin, and Blaed, standing beside an old weather-beaten barn overlooking the ocean. Michael stood slightly apart from the others, gripping a blanket around his shoulders and occasionally retching. Tony asked how my night was going, and I tried to describe my visions. Classic DMT hallucinations, Kevin said.

Blaed said the ayahuasca's effects were interesting, but he had expected something more intense and long-lasting, more like LSD. Tony said Blaed should have had another supplement. He told us he recently supervised an ayahuasca session with a Norwegian psychotherapist who couldn't get enough of the stuff, even though it made him very sick. After vomiting for the umpteenth time, the therapist crawled over to Tony on his hands and knees and groaned, "More." Everyone laughed at Tony's story.

All at once I felt the same peculiar combination of weakness and surging energy that had signaled the onset of the ayahuasca. Dizzy, I drifted away from the group and stared at the ocean. The moon seemed small, impossibly distant. Milky light poured down from it and vaporized as it struck the ocean, which was black and luminous, like molten lead, sheathed in a silvery miasma.

Something shifted, and the scene turned horrific. The moon was the sun, shrunken to a white dwarf, its life-sustaining heat and energy long since radiated away, barely illuminating the chill cinder of the earth. I was seeing the future, long after humanity and all of life had vanished from the planet. The flame of consciousness had flickered out in the eternally expanding cosmos, and it had reverted to dumb, blind, painless, meaningless matter, as it must.

Untying the rope

Michael began retching again. Kevin said we should probably head back and see how the others were doing. Shuffling toward the ranch with my companions, I felt stunned, disoriented, isolated by my end-of-life epiphany. Several times I opened my mouth to tell the others about it, but the words didn't come. Gradually, my comrades' conversation dragged me back from the lifeless future to the present.

Blaed, the youngster among us, complained about Tony's music selections, especially that Tibetan shit. Blaed's uncle Kevin agreed that the Tibetan music was awful. I realized they meant the droning dirge that drove me from the circle. I said nothing, worried that Tony's feelings might be hurt, but Tony accepted the criticism with good humor and promised to rethink his selections. Blaed rattled off the names of several groups whose music would be great for an ayahuasca session. None of the names was familiar to me. I suggested that some old psychedelic rock might be nice, like Iron Butterfly's "In-A-Gadda-Da-Vida." My companions laughed, apparently thinking I was joking.

Did anyone see that meteor earlier tonight? Blaed asked. He'd never seen a meteor so big and so close; he saw burning chunks splitting off of it! Blaed and Kevin took turns pointing out different constellations. Someone asked why the Pleiades looked so misty. It's a huge cloud of gas, I said, still in the process of condensing into stars. Astronomers call it a stellar nursery.

That set Tony off on a discussion of the ubiquity of birth metaphors in science and in the myths of indigenous peoples. He was intrigued by the theory of Stanislav Grof that the trauma of birth plays a crucial role in shaping our psyches but thought Grof tried too hard to link all transpersonal experiences to perinatal events. No single theory, Tony said, can explain the immense variety of psychedelic visions.

We arrived back at the circle, around which sleeping bags sprawled like monstrous larvae. Exhausted, I climbed into my bag and dozed off. When my eyes opened the sky was brightening, the stars fading. I walked away from the circle to write down some notes. Michael, who had been too sick to speak a few hours ago, sat beside me. He was feeling much better. He had only recently become interested in psychedelics, he told me. His experiences helped him cherish his family, his wife and two-year-old daughter, and they reminded him that there's more to life than the rat race. His wife was a devout Christian and fairly conservative; she didn't join him in taking psychedelics, but she understood why he did.

When we returned to the circle, everyone was awake, comparing notes. Blaed asked whether anyone else heard or saw strange flashes, booms, and howling noises during the night, or was it just him? Others confirmed that they'd heard the same noises. Actually, said Linda, the explosions were real; someone at a nearby ranch was apparently setting off fireworks. As for the howling, she added, those were coyotes. If only it were always so easy to distinguish reality from illusion, I thought.

Nancy, the muscular firefighter, was beaming. She'd had a cathartic trip in which she relived lots of things from her childhood; the only bad part was that she threw up on herself. George said his experience was disappointing; he vomited early and often and never had any hallucinations. You should have asked for a supplement, Tony chided him.

This was a pretty uneventful night, Tony said. Usually at least one person becomes convinced he's going insane or dying. Tony had learned tricks from shamans that can help people pass through these ordeals. He hums and lays his hands on the sufferer's forehead or solar plexus, or he blows between the person's shoulder blades or on the top of the head. He had no idea why these tricks work, but they do.

We walked to the main house for a breakfast of bagels, scrambled eggs, and coffee. Tony, seated beside me, asked me how I felt. Surprisingly good, considering, I replied. That's typical, Tony said. People often feel refreshed the morning after an ayahuasca session. It can be a wonderful time for writing, painting, composing music, doing anything creative. He hoped someday to set up a center where scientists, artists, businesspeople, and others could take ayahuasca for creative problem-solving. I tried to imagine corporate executives pondering Internet marketing strategies as they spewed out Day-Glo vomit.*

After breakfast, we headed back to the circle for a final ceremony. Each of us was supposed to untie the knot we made in Tony's rope the previous evening and say something. For the most part, everyone stuck to

---

*Tony's suggestion that psychedelics be used at corporate retreats may not be so crazy after all. Many young businesspeople already use psychedelics for creative problem-solving, according to "Brave New Work," by Michael Shrage (*Fortune*, April 30, 2001). Shrage suggested that these productive psychedelic enthusiasts could help bring about reforms in the drug laws.

fairly generic expressions of thanks — to Allen and Deborah for letting us use their home, to Tony and Kevin for supplying the ayahuasca and guiding us, to the entire group for just being there, and to the ayahuasca plant spirits for transporting us into their realm. The last person to speak was Deborah, who seemed so melancholy the day before and had been quiet all morning. She said a few words, haltingly, then broke down sobbing. Allen took her in his arms and rocked her back and forth. Trying to lighten the mood, Kevin asked: If God takes LSD, does He see people? To my ears, the joke sounded more ominous than funny.

It was almost noon when I said goodbye to my companions. I drove a few miles down the road to an inn, deep in a redwood forest, where I had reserved a room. Lying on the bed with a notebook in my lap, I tried to recollect as much as I could from the previous evening and to make sense of it all. Considering how long it had been since I had slept, my mind felt strangely clear, almost too clear.

There was something both touching and absurd about the whole session. We went to great lengths to take this stuff that we knew would sicken us and might frighten us half to death. Why? Perhaps for the same reason we watch horror movies: We expose ourselves to this simulacrum of darkness to inoculate ourselves against the true darkness that will claim us all someday.

In retrospect, all my ayahuasca visions seemed more like products of my own brain than transpersonal revelations. Ayahuasca showed me a universe of dazzling, alien pyrotechnics. As our group leader Kevin told me, those are characteristics of DMT, just as sensations of communion and compassion are characteristics of MDMA. Maybe the philosopher Steven Katz was right: When we take a drug, we don't discover reality, we just discover a drug.

The visual hallucinations reminded me of the neurophysiological theory of form constants, the geometrical patterns caused by psychedelics. Each time a neuron discharged in my visual cortex, it triggered a cascade of neural firing around it according to the chemical rules dictated by DMT, harmine, and harmaline, ayahuasca's active ingredients. Subjectively, I perceived these effects as a wildly convecting cellular automaton — albeit one with infinite dimensions and hues. And perhaps because of some miscommunication between my brain's hemispheres, I attributed the visual hallucinations to an alien intelligence, or what the psychologist Michael Persinger would call a sensed presence.

As for the end-of-life vision I had while gazing out over the Pacific, well, that was just a waking version of nightmares I'd had since childhood, when I first learned that missiles could fall from the sky at any moment and destroy everything. That was not to say my psychedelic vision was wrong. The laws of probability dictate that now and then our dreams and nightmares come true.

I had hoped that ayahuasca would give me some insight that I could take back to my wife to comfort her. But what could I say? That I glimpsed a cold, inhuman intelligence underlying reality? A Demiurge without a heart? That I foresaw the end of the world, of life, of consciousness? Real comforting. I preferred the anxious, all too human Demiurge that I saw — or rather, became — during my 1981 drug trip. At least that was a God I could identify with, who helped me make sense of the world's imperfections. Of course, that God, like all gods, is just an illusion, a product of the innate human compulsion to find order in the world, an anthropomorphic order. If we look hard enough, we can see our own reflections everywhere.

Trying to sum things up, I wrote: "So, what is the final lesson? Do I see only what I already believe? Is it too late to learn? Change? Do I want to learn and change? Need to?" I stared at the words, wracking my brain for an epiphany, but nothing came. I felt more confused, more mystified than ever. What is the point of all our mystical searching? Where does it get us? What truth does it give us? What consolation?

It was late afternoon when I pulled on shorts and jogging shoes and ran down the dirt road leading away from the inn. As I huffed and puffed up a hill lined with stoic redwoods, a station wagon pulled up next to me. At the wheel was Deborah. The last time I saw her, just after the rope ceremony that morning, she had been puffy-eyed, submerged in her secret grief. She seemed composed now, her aura of sadness dissipated. Beside her was a boy, perhaps ten years old, who had her fair hair and skin and wide-set blue eyes. Her son, Deborah said, introducing us. She said she was glad to have met me, and I said likewise. Still catching my breath, I stood and watched as her car climbed slowly up the hill and vanished around a curve.

12

The Awe-ful Truth

Greeting me at the door of his home in Berkeley, Huston Smith seemed almost ethereal, frailer than the last time I had seen him, more than a year earlier. He moved, spoke, and met my eyes with calm deliberation, like a man carrying a glass filled to the brim with water. After getting take-out food from a Chinese restaurant near his house, we sat down for lunch in his dining room. A vase of pink flowers stood on the table between us. Over Smith's shoulder I could see a bronze statue of Buddha, with his half-lidded eyes and maddening Mona Lisa smile: I'm enlightened, and you're not. Smith had helped me find my bearings at the start of this mystical inquest, and I thought that he, more than anyone else I had spoken to, might help me reach some sort of resolution.

I told Smith that since he had explained the perennial philosophy to me in Albuquerque, I had consulted many other authorities on mysticism, and they had left me more mystified than ever. They disagreed about the causes of mystical experiences, and their meaning. They had different notions of enlightenment, if they believed in it at all. Some mystics envisioned a loving, all-powerful God; others postulated a flawed God or rejected the concept of God outright. I added that a few months ago I had taken an entheogen, ayahuasca, that had compounded my confusion. As I vented, Smith nodded and smiled, as if to say, Welcome to my world.

I wanted to believe him and other perennialists who said that mysticism reveals a universal truth, I said, but that truth kept slipping from my grasp. For example, does the perennial philosophy insist that consciousness is eternal or that there is life after death? "I would hope it doesn't *insist* on anything," Smith replied. The wisdom traditions, he said, all proclaim in one way or another that our consciousness is not just an epiphenomenon of the brain extinguished at death; it is more enduring than materialistic science implies.

But if he had to boil the perennial philosophy down to a single tenet, Smith continued, it would be this: Beyond our mundane, material world lies an infinite, transcendent realm. Tapping his plate with his chopsticks, he asked me to imagine that it is "the infinite," which is "that outside of which it is impossible to fall"; the sesame noodles on the plate represent the finite realm in which we live. A mystical experience "makes the obtuse blockages to the infinite transparent, so you can look through the noodles" — he jabbed the glutinous tangle with his chopsticks — "to see the infinite" — he tapped the rim of his plate.

But isn't even this tenet debatable? I asked. Certain mystics have insisted that they aren't transported to some transcendent "beyond"; they just see ordinary reality for what it really is. Nodding, Smith responded with an old Zen saying: "Before I began my practice, mountains were mountains, and rivers were rivers. In the middle of my practice, mountains were no longer mountains, and rivers were not rivers. But now that I've completed my practice, mountains are mountains, and rivers are rivers."

Are the first mountains and rivers the same as the last? Smith asked, adding, "Thirty blows if you answer! Thirty blows if you don't!" The world hasn't changed, he explained, but our *perception* of it certainly has. Mystical awakening infuses our vision with awe, which "may be *the* distinctive religious emotion." As the German theologian Rudolf Otto said, in his book *The Idea of the Holy*, awe "combines fear and fascination," Smith said. "Fascination because new realms of being are opening to you that you didn't even know were there, and what could be more fascinating? And fear because this is unknown territory, and are the natives friendly?"

At the time, Smith's formulation of mysticism — awestruck perception of the infinite in the finite — seemed frustratingly vague. It made more sense after I returned home and read *The Idea of the Holy*, as Smith had recommended. The power of Otto's book derives in part from its gothic obscurity. Otto preferred to use the Latin terms *numen* and *mysterium tremendum* rather than *God* to describe the inner sanctum of reality, that which we confront in a mystical vision. Our encounter with the *mysterium tremendum* can strike us "chill and numb" and fill us with "an almost grisly horror and shuddering" (a state that could be called mysterium tremens).

Lurking within Otto's dank prose was a radical thesis, especially given that its author was a Christian. It reminded me of the negative theology I had been exposed to at the "Mystics" conference in Chicago. The *mysterium tremendum,* Otto said, is not anything we can possibly identify with, let alone become. It is not a deity, force, principle, spirit, or ground of being — not a thing at all. It is "wholly other," "nothingness," the "opposite of everything that is and can be thought." It is absence, not presence.

Religions, Otto argued, do not reveal the *mysterium tremendum* so much as they shield us from direct confrontation with it. "They are attempts . . . to guess the riddle it propounds, and their effect is at the same time always to weaken and deaden the experience itself." Theology "often ends by constructing such a massive structure of theory and such a plausible fabric of interpretation, that the 'mystery' is frankly excluded." Otto derided primitive mythologies in particular — crowded as they are with gods and goddesses and spirits — as "ghost stories" that divert us from the one true *mysterium tremendum.*

I decided that Huston Smith had pinpointed the essence of mysticism after all. He had also helped me rediscover a truth I had intuited in my own entheogenic experiences. What Smith called "the infinite" is what Otto called the *mysterium tremendum,* the *numen,* the "wholly other." It is the nothingness from which we came and to which we must return. In one way or another, all mystics intuit the infinite, whether they call it Yahweh, God, Allah, ground of being, Brahman, void, or *mysterium tremendum.* There is no predicting how you will respond to such a vision. Seeing life against the backdrop of infinity can evoke joy, madness, terror, revulsion, love, gratitude, hilarity — or all of the above at once. You may delight in the world's astonishing beauty or despair at its fragility and insignificance.

You may also feel compelled to "guess the riddle," to explain how our finite human something could have emerged from that infinite inhuman nothing. But this is the one riddle that cannot be solved. A paradox arises. I said at the beginning of this book that a mystical experience must have a noetic quality; you must believe you are seeing "the Way Things Are." Mystical awe is the inverse of knowledge; it is a kind of anti-knowledge. Instead of seeing The Answer to the riddle of existence, you see just how impenetrable the riddle is.*

---

*No modern spiritual writer emphasizes awe and doubt more than the British Buddhist Stephen Batchelor. He advocates an "agnostic Buddhism," which seeks to cultivate perplexity

At the same conference where I first met Huston Smith, the former astronaut Edgar Mitchell described an unusually literal mystical experience. Hurtling toward the moon on the *Apollo* 14 mission in 1971, Mitchell peered out the window of his spacecraft and saw what Smith had tried to represent with his sesame noodles and plate: our tiny human world adrift in boundless space. Mitchell's initial reaction was overwhelming awe. Upon reflection, he became convinced that mainstream, materialistic science cannot account for our existence. He was right. But Mitchell was wrong to think that the *mysterium tremendum* can be solved by a science souped up with New Age physics and paranormal effects. His attempt to "guess the riddle" was just another ghost story. So was my "gnosticism lite" theology, which was inspired by my entheogenic moon shot in 1981.

Our compulsion to explain the *mysterium tremendum* is understandable. As Rudolf Otto warned, confronting the abyss of nonbeing can be terrifying. To the extent that you realize life's improbability, you realize its precariousness and gratuitousness. You know that there is no reason for us to exist. The odds seem overwhelming that our minuscule human hubbub will be swallowed up by the emptiness whence we came. But the flip side of mystical terror is joy. We should not be here, and yet here we are. How lucky can we get? In other words, heaven and hell are two sides of the same mystical vision.

Rather than asking who is mystical, it might be better to ask who is anti-mystical. You are anti-mystical, I propose, to the extent that you think reality has been or can be explained, and I mean really *explained*, by Hinduism or theosophy or gnosticism lite or superstring theory or any other theory or theology. Mystics can be anti-mystical, to the extent that they think their visions have revealed The Truth, The Answer, the secret of life. Ken Wilber, when he implies that his transpersonal periodic table can account for every aspect of reality, is anti-mystical. But even Wilber, as brimming with certainty as he is, has written that at the heart of things is "a staggering mystery" that "facts alone can never begin to fill."

---

before the mystery of existence. This state is not always pleasant. When we truly confront reality, we "tremble on that fine line between exhilaration and dread," Batchelor wrote in *Buddhism Without Beliefs* (Riverhead Books, New York, 1997, p. 102). Batchelor was inspired to follow this path by an incident that occurred in 1980, when he was at a monastery in India. Carrying a bucket of water to his hut, he suddenly stopped short, overwhelmed by a sense of the mystery of existence. This epiphany, which lasted only a few minutes, was not "an illumination in which some final, mystical truth became momentarily very clear," he recalled in *The Faith to Doubt* (Parallax Press, Berkeley, California, 1990, p. 10). "For me it gave no answers. It only revealed the massiveness of the question."

Most of the other mystical experts I interviewed espoused awe-ful worldviews, if only implicitly. The anti-perennialist philosopher Steven Katz said that "ultimate reality, by its very nature, escapes us." The neurologist James Austin assured me that our existence is "beyond belief." Stanislav Grof acknowledged that when it comes to creation, "we are stuck with a mystery that you cannot account for." Terence McKenna knew that gnosis can never capture the weirdness of the world. Here is where many different paths converge: The perennial philosophy, postmodernism, negative theology, transpersonal psychology, neurotheology, gnosticism, and neo-shamanism all insist in their own ways that there is an irreducible mystery at the heart of things. So does science.

Science and Mysticism Reconciled!

There are many claimed convergences between science and mysticism. Cognitive psychology supposedly corroborates the Buddhist doctrine that the self is an illusion. Quantum mechanics, which implies that the outcome of certain microevents depends on how we measure them, is said to confirm the mystical intuition that consciousness is an intrinsic part of reality. Similarly, quantum nonlocality, which Einstein disparaged as "spooky action at a distance," clinches mystics' perception of the interrelatedness, or unity, of all things. I see a different point of convergence between science and mysticism: Each in its own way reveals the miraculousness of our existence.

The more science learns about the origin and history of the cosmos and of life on earth and of *Homo sapiens,* the more it reveals how staggeringly improbable we are. First there is the fact of existence itself. The big bang theory represents a profound insight into the history and structure of the cosmos, but it cannot tell us why creation occurred in the first place. Particle physics suggests that empty space is seething with "virtual particles," which spring into existence for an instant before vanishing. In the same way, some physicists speculate, the entire universe might have begun as a kind of virtual particle. Honest physicists will admit that they have no idea why there is something rather than nothing. After all, what produced the quantum forces that supposedly made creation possible? "No one is certain what happened before the Big Bang, or even if the question has any meaning," Steven Weinberg, the physicist and advocate of the nature-does-not-care principle, wrote recently.

Next questions: Why does the universe look this way rather than some other way? Why does it adhere to these laws of nature rather than to some other laws? Altering any of the universe's fundamental parameters would have radically altered reality. For example, if the cosmos had been slightly more dense at its inception, it would have quickly collapsed into a black hole. A smidgen less dense, and it would have flown apart so fast that there would have been no chance for stars, galaxies, and planets to form. Cosmologists sometimes call this the fine-tuning problem, or, more colorfully, the Goldilocks dilemma: How did the density of the universe turn out not too high, not two low, but just right?

The odds that matter would have precisely its observed density, the physicist Lawrence Krauss has calculated, are as great as the odds of guessing precisely how many atoms there are in the sun. Some physicists are so troubled by the arbitrariness of the cosmos that they espouse a quasi-theological concept known as the anthropic principle. According to this notion, the universe must have the structure we observe, because otherwise we wouldn't be here to observe it. The anthropic principle is cosmology's version of creationism.*

The next improbability is life. The evolutionary biologist Richard Dawkins once declared that life "is a mystery no longer," because Darwin solved it with his theory of evolution by natural selection. Yet life is as mysterious as ever, in spite of all the insights provided by evolutionary theory and more recent biological paradigms, such as genetics and molecular biology. Neither Darwinism nor any other scientific theory tells us why life appeared on earth in the first place or whether it was probable or a once-in-eternity fluke.

Many scientists have argued that life must be a ubiquitous phenomenon that pervades the universe, but they can offer precious little empirical

---

*The anthropic principle comes in two forms, weak and strong. The weak anthropic principle, or WAP, holds merely that any cosmic observer will observe conditions, at least locally, that make the observer's existence possible. The strong anthropic principle, SAP, insists that the universe *must* be constructed in such a way so as to make observers possible. WAP is tautological and SAP teleological. The physicist Tony Rothman, a former colleague of mine at *Scientific American*, liked to say that the anthropic principle in any form is completely ridiculous, and hence should be called CRAP.

evidence to support this assertion. After decades of searching, scientists have found no signs of life elsewhere in the cosmos; a 1996 report of fossilized microbes in a meteorite from Mars turned out to be erroneous. Researchers still cannot make matter animate in the laboratory, even with all the tools of biotechnology. In fact, the more scientists ponder life's origin, the harder it is to imagine how it occurred. Francis Crick once stated that "the origin of life appears to be almost a miracle, so many are the conditions which would have to be satisfied to get it going." In his book *Life Itself,* Crick offered the McKenna-esque speculation that the seeds of life might have been planted on earth by an alien civilization.

Once life on earth started evolving, many scientists have contended, it was only a matter of time before natural selection produced a species as intelligent as *Homo sapiens.* But for more than 80 percent of life's 3.5-billion-year history, the earth's biota consisted entirely of single-celled organisms such as bacteria and algae, so not even the simplest multicellular organisms were inevitable. The evolutionary biologist Stephen Jay Gould has estimated that if the great experiment of life were rerun a million times over, chances are that it would never again give rise to mammals, let alone mammals intelligent enough to invent negative theology and television. Similar reasoning led the eminent evolutionary theorist Ernst Mayr to conclude that SETI — the search for extraterrestrial intelligence program, which scans the heavens for radio signals from other civilizations — is futile.

Multiply all these improbabilities and they spike to infinity. As the mystical goat Susan Blackmore has pointed out, we are bad at judging probabilities and hence prone to making too much of chance events; that is why we believe in ESP, clairvoyance, telekinesis, and other miracles. I do not believe in miracles, at least not defined in the conventional religious manner as divine disruptions of the natural order. But if a miracle is defined as an infinitely improbable phenomenon, then our existence is a miracle, which no theory natural or supernatural will ever explain.

Scientists may go much further in plumbing nature's secrets. They may decipher the neural code, the secret language of the brain. They may arrive at a plausible explanation of how life emerged on earth, and they may discover life elsewhere in the cosmos. They may find and verify a unified theory of physics, which will provide a more precise picture of the origin and history of the universe. Although there are good reasons for doubting such scientific advances, they cannot be ruled out. What can be ruled out is that science will answer the ultimate question: How did something come from nothing? Neither superstring theory nor any other of science's so-called theories of everything can resolve that mystery, any more than our supernatural theologies can.

Some researchers, notably the British chemist Peter Atkins, have proclaimed that science will soon solve the riddle of existence once and for all. Atkins is, if possible, more adamant an atheist and a religion-basher than his countryman Richard Dawkins. More than a decade before I argued, in *The End of Science,* that science's grand quest to uncover the basic rules governing reality might be reaching an impasse, Atkins wrote: "Fundamental science may be almost at an end, and might be completed within a generation." Whereas I saw science leaving many riddles unresolved, Atkins envisioned science as a universal acid dissolving all mysteries, including spiritual, ethical, and philosophical ones. "Complete knowledge is just within our grasp," he intoned. "Comprehension is moving across the face of the earth, like the sunrise."

I suspected Atkins of rhetorical excess until I met him at a reception in London in 1997. As we sipped white wine from plastic cups, Atkins assured me with almost demonic glee that science would soon provide such satisfying explanations for all of nature's mysteries — including the supreme riddle of the universe's origin — that wonder would be extinguished forever, and along with it all of our silly spiritual superstitions. The world would be as solved as Fermat's last theorem: QED.

It is anti-mystics like Atkins who provoke Huston Smith's distrust of science, or rather, scientism. But Atkins is an extremist, on the fringe of the atheistic fringe. Even hard-core scientific materialists generally acknowledge that science can never — will never — dispel our awe before the mystery of existence. In fact, science relies on awe for inspiration at least as much as religion does. "Our sense of wonder grows exponentially: the greater the knowledge, the deeper the mystery," wrote Edward O. Wilson. "This catalytic reaction, seemingly an inborn human trait, draws us perpetually forward in a search for new places and new life." In this shared awe, perhaps, there is hope for a rapprochement between science and spirituality.

The astronomer and author Chet Raymo has offered five compelling reasons for embracing the "New Story" of science rather than the old stories of religion as the basis for spirituality: One, science works. Two, it is universal, true for all people at all times. Three, it emphasizes the connectedness of all people and all things. Four, it makes us, rather than some deity or transcendent force, responsible for our own destiny. Five, it reveals the universe to be more complex, vast, and beautiful than we ever imagined. But what the New Story of science cannot do, Raymo said, is guarantee our survival. "We are contingent, ephemeral — animated stardust cast up on a random shore, a brief incandescence," he wrote. "The New Story makes one thing clear: *We are not immortal.*"

Huston Smith sees this as one of science's chief deficiencies, that "there is no way that a happy ending can be worked into it." To my mind, the phrase *happy ending* is a contradiction in terms; the only happy ending is no ending. I certainly don't see the Last Judgment as a happy ending, nor the Buddhaverse. I prefer theologies such as Terence McKenna's timewave and Freeman Dyson's principle of maximum diversity, which offer endless adventure rather than closure. Although science cannot promise us immortality, neither can religion. Moreover, scientists have imagined ways in which consciousness might last forever, not through divine intervention but through our own ingenuity.

Dyson has calculated that consciousness — albeit in the form of clouds of charged particles rather than flesh and blood — might resist entropy and sustain itself forever in an eternally expanding universe through shrewd conservation of energy. Dyson was piqued into making these calculations by Steven Weinberg's comment: "The more the universe seems comprehensible, the more it also seems pointless." Dyson retorted, in effect, that no universe with conscious life is pointless. Other scientists have imagined how consciousness might endure even if the universe eventually stops expanding and collapses into an infinitely dense "Omega Point." I call suppositions such as these "scientific theology," because they are little more than theology — speculation about ultimate ends — clad in flimsy scientific garb. Scientific theology nonetheless demonstrates that science can imagine futures at least as hopeful and open-ended as those of religion.

Art, Irony, and Garbage

Although we can never solve the riddle of existence, we can never stop trying. We must keep reimagining our relationship to the infinite. Skepticism alone — and the cold, hard facts of science — cannot serve as the basis for spirituality. Susan Blackmore helped me reach this conclusion. She described Zen as a kind of rubbish-removal system that cleanses the mind of extraneous beliefs and emotions so that we can see reality as it truly is.

I found Blackmore's garbage metaphor appealing at first, because it provided a handy criterion for judging theories and theologies. The worst ones, I decided, distract us from the reality right in front of us by postulating parallel dimensions and universes, heavens and hells, gods and ghosts and demiurges and extraterrestrials. Too much garbage! Viewed this way, skepticism appears to be the ideal spiritual perspective. Skepticism clears away cumbersome beliefs on an intellectual level, just as meditation (ideally) clears away beliefs, emotions, and thoughts on a more experiential level. Skepticism can help us achieve mystical deautomatization, or so I wanted to believe.

My handling of real rather than metaphorical garbage gradually gave me a more complicated view of the matter. In my kitchen, we put garbage in bags that come in boxes of twenty. After I yank the last bag from its box, the box becomes trash, which I put in the bag. Sometime after I interviewed Susan Blackmore, every time I pulled the last bag from the box and stuffed the box in the bag, I intuited a paradox lurking within this ritual.

I went through more garbage bags than I care to mention before I solved the riddle: Every garbage-removal system generates garbage. Zen apparently works as an efficient garbage-removal system for Susan Blackmore and James Austin. But as minimalistic as it is, Zen clutters more than it clarifies my mind.* Once I started down this line of thinking, it was hard to stop. I began looking askance at skepticism, too. Maybe skepticism, instead of cleansing our vision, just substitutes one type of trash for another. Instead of belief in reincarnation, angels, ESP, E.T., parallel universes, and the

---

* I took classes in Zen in my hometown in 1999–2000, but my efforts to become what my wife calls a "koan-head" failed. My doubts about Buddhism have been reinforced by one of my favorite books, *The Snow Leopard,* by Peter Matthiessen (Bantam Books, New York, 1979), which chronicles Matthiessen's trek into the Himalayas in search of the elusive snow leopard. Matthiessen, whose wife died of cancer shortly before his journey, was in search of spiritual solace, perhaps even enlightenment. He recounted not only external events but also inner, psychic ones, which he often interpreted in Buddhist terms. He kept agonizing over his inability to achieve pure, ecstatic perception, or mindfulness. It struck me eventually that Matthiessen's concerns about mindfulness were preventing him from achieving it. His gruff partner, the zoologist George Schaller, was probably more mindful because he did not give a whit about mindfulness or any other Buddhist nonsense. He just went about his business.

Oedipus complex, the skeptic crams his mind with *disbelief* in reincarnation, angels, and so on.

The problem is that any truth or anti-truth, no matter how initially revelatory and awe-inspiring, sooner or later turns into garbage that occludes our vision of the living world. Ludwig Wittgenstein had this problem in mind when he described his philosophy as a ladder that we should "throw away" after we have climbed it. At its best, art — by which I mean poetry, literature, music, movies, painting, sculpture — works in this manner. Art, the lie that tells the truth, is intrinsically ironic. Like Wittgenstein's ladder, it helps us get to another level and then falls away. What better way to approach the mystical, the truth that cannot be told?*

At the "Mystics" meeting in Chicago, one speaker warned that if we can't talk about mysticism, we can't whistle about it, either. In other words, all our modes of expression, including art, fall short of mystical truth. But unlike more literal modes of expression, art comes closer to uttering the unutterable by acknowledging its own insufficiency. It gives us not answers but questions. That does not mean mystical insights cannot be expressed within other modes of knowledge, such as science, philosophy, theology — and of course journalism. But we should view even the most fact-laden mystical texts ironically when they turn to ultimate questions. Some mystical writers, notably Terence McKenna, supply their own irony, but we readers can supply it even if the author intended none. We can read the Upanishads, Genesis, Dionysius the Areopagite, and the neurotheological suppositions of Andrew Newberg as we read Blake or Borges or Emily Dickinson.

Viewed ironically, even the most fantastical ghost stories, including the old stories of religion, can serve a purpose. Whether they postulate superintelligent clouds of gas, insectoid aliens in hyperspace, a Demiurge with multiple-personality disorder, or a loving God who for inscrutable reasons makes us suffer, well-told ghost stories can remind us of the unfathomable mystery at the heart of things. Our creation myths and escha-

---

*Joseph Campbell made this point in *The Masks of God* (Penguin Books, New York, 1978, p. 3): "Prayers and chants, images, temples, gods, sages, definitions, and cosmologies are but ferries to a shore of experience beyond the categories of thought, to be abandoned on arrival." Campbell suggested that poetry is the best medium for spiritual expression. When poets take their own imagery too literally, Campbell warned, they degenerate into prophets — or, worst of all, priests, tenders of dogma (pp. 518–519).

tologies, our imaginings of ultimate beginnings and ends, can also help us discover our deepest fears and desires. But sophisticated scientific theology should never be mistaken for ultimate truth. What Voltaire said centuries ago still holds, and will always hold: "It is truly extravagant to define God, angels, and minds, and to know precisely why God defined the world, when we do not know why we move our arms at will. Doubt is not a very agreeable state, but certainty is a ridiculous one."

The problem of fun

Other than art, is there any method particularly suited for evoking mystical awe without the side effects that so often attend it? In *Psychedelic Drugs Reconsidered,* Lester Grinspoon and James Bakalar suggested that the chief benefit of psychedelics is "enriching the wonder of normality" — that is, enhancing our appreciation of ordinary consciousness and ordinary life. This is the spiritual value cited most often by entheogenic advocates such as Huston Smith. But psychedelics can have the opposite effect. This world may seem drab in comparison to the bizarre virtual realms into which LSD or DMT propels us. Instead of opening our eyes to the miraculousness of everyday reality and consciousness, psychedelics can blind us.

All mystical technologies that induce powerful altered states pose this risk. One mystical expert who has reached this conclusion is Jean Houston. A pioneer of the human potential movement, Houston works as a kind of spiritual psychotherapist, usually for large groups rather than individuals. She seeks to rejuvenate her clients' psyches through dance, song, chanting, guided imagery, and role-playing, often with a mythological dimension. She and her husband, the anthropologist Robert Masters, proclaimed in 1966 that investigations of LSD and similar drugs could help human consciousness expand "beyond its present limitations and on towards capacities not yet realized and perhaps undreamed of."

Houston subsequently became quite critical of the *via psychedelica.* "I am by nature not pro-drug," she told me. Timothy Leary was one of the most charming people Houston had ever met — and one of the most irresponsible. Too many people lured onto the psychedelic path by this Pied Piper suffered breakdowns and ended up in mental hospitals, Houston said. "If I were to take the American pragmatic tradition and say, 'By their fruits ye shall know them,' then I'd have to say I haven't seen too much evidence" that psychedelics promote a healthy spirituality. "Some might say it is a shortcut to reality. But the fact is, it doesn't seem to sustain that reality."

Houston's disillusionment with psychedelics led her to seek safer means of self-transcendence. In the early 1970s, she and Robert Masters invented what they called the Altered States of Consciousness Induction Device, or ASCID. It consisted of a suspension harness in which blindfolded subjects could spin around in three dimensions. The contraption worked so well that Houston and Masters discontinued its use. "People would get addicted to it and even refuse to explore their inner states without first taking a ride," Houston recalled. The experience reinforced her suspicion that any spiritual practice or path — and particularly those emphasizing altered states — can become an end in itself, which leads us away from reality rather than toward it.*

Anything that helps you see — really *see* — the wondrousness of the world serves a mystical purpose. According to Zen legend, when a visitor asked the fifteenth-century master Ikkyu to write down a maxim of "the highest wisdom," Ikkyu wrote one word: "Attention." The visitor asked, irritably, "Is that all?" This time, Ikkyu wrote two words: "Attention. Attention." Fortunately, life itself is so wildly weird and improbable that sooner or later it is bound to get our attention. And if life doesn't grab our attention, death will. Whenever death intrudes upon our lives, we feel the chill of the deep space in which we are suspended.

Spiritual seekers have employed memento moris, such as a human skull, to keep themselves mindful of death. An extreme version of this technique, used in certain Buddhist sects, involves sitting next to or on top of a rotting corpse. It seems that this practice may merely desensitize you to death rather than sensitizing you to life. Moreover, dwelling on death, the abyss, nothingness, may convince you that it is the only abiding reality and that all finite, time-bound phenomena, including our mortal selves, are

---

*One indication of how desperate people can become in search of new-and-improved mystical technologies is the recent interest in trepanation, which calls for drilling a small hole in the skull. While attending a conference titled "Altered States of Consciousness," held at New School University in New York, February 22–24, 2001, I met Amanda Fielding, a British woman who has been proselytizing for trepanation ever since drilling a hole in her own forehead in 1970. A movie that Fielding made of her self-trepanation, called *Heartbeat in the Brain*, has become a cult film. Fielding, who is also known as Lady Neidpath, promotes trepanation through the Trepanation Trust (www.trepanation.com).

ephemeral and hence in some sense unreal.* To be enlightened, Ken Wilber once wrote, is "to snap out of the movie of life." This is perhaps the greatest danger posed by mysticism, that you will be left with a permanent case of derealization and depersonalization.

If you are lucky, your glimpse of the abyss will make this life seem more real, not less. You will feel what Albert Hofmann felt after the psilocybin trip in which he had found himself all alone in a ghost town inside the earth. When he returned from this hellish solitude, back to the world and his dear friends, he felt "reborn," and he was overcome with gratitude and joy at the "wonderful life we have here." This is by far the greatest gift that mystical experiences can bestow on us: to see — really *see* — all that is right with the world. Just as believers in a beneficent deity should be haunted by the problem of natural evil, so gnostics, atheists, pessimists, and nihilists should be haunted by the problem of friendship, love, beauty, truth, humor, compassion, fun.† Never forget the problem of fun.

---

*Even Huston Smith has used rhetoric that exalts the infinite at the expense of quotidian reality. In *Cleansing the Doors of Perception* (Jeremy Tarcher/Putnam, New York, 2000, p. 133), Smith said the "basic message of the entheogens" is that there is "another Reality that puts this one in the shade." In *Mysticism Sacred and Profane* (Oxford University Press, New York, 1961), the British theologian R. C. Zaehner denigrated "nature mysticism," in which the mystic embraces the material world, as inferior to union with a totally transcendent God. Zaehner approvingly quoted the Sufi master Ghazali saying that love of God and love of the world are "mutually exclusive" (p. 61). It was no doubt to distinguish Zen from these views that D. T. Suzuki once called Zen "radical realism rather than mysticism" (as quoted in *Zen and the Brain,* by James Austin, MIT Press, Cambridge, 1998, p. 16).

†Some evolutionary biologists have taken on the problem of beauty. In *Climbing Mount Improbable,* W. W. Norton, New York, 1996, p. 236, Richard Dawkins recalled asking his six-year-old daughter, who had just pointed out wildflowers they were passing in their car, what she thought wildflowers are for. "To make the world pretty, and to help the bees make honey for us," she replied. "I was touched by this," Dawkins wrote, "and sorry I had to tell her it wasn't so." He argues, quite rightly, that this kind of anthropocentric thinking leads Christian fundamentalists to claim that God created the AIDS virus to punish sinners. But Dawkins never really explains why so much of nature seems so extraordinarily beautiful to us. Edward O. Wilson has suggested that natural selection may have instilled in us a "biophilia," or reverence for nature, that benefits both us and those creatures with whom we enjoy mutually beneficial relationships. But why do we respond to rainbows, sunsets, and stars, phenomena from which we extract no tangible, utilitarian benefit? As Steven Weinberg wrote in *Dreams of a Final Theory* (Pantheon, New York, 1992, p. 250), "I have to admit that sometimes nature seems more beautiful than strictly necessary."

Rejecting the Oneness and Enlightenment Memes

To some mystics, awe, wonder, and astonishment are just side effects of mystical awareness.* They often tout oneness rather than wonder as the essence of mysticism. In *The Varieties of Religious Experience,* William James singled out oneness — "the overcoming of all the usual barriers between the individual and the Absolute" — as the dominant theme in the writings of all mystics. "In Hinduism, in Neoplatonism, in Sufism, in Christian mysticism, in Whitmanism, we find the same recurring note, so there is about mystical utterances an eternal unanimity which ought to make a critic stop and think."

But perhaps oneness-worshipers should stop and think. They might also read a critique of oneness and related doctrines by Diana Alstad and Joel Kramer, two veterans of the American alternative-spirituality scene. Alstad and Kramer have taught workshops on yoga and male-female relations since the early 1970s, but eventually they became disillusioned with Buddhism, Hinduism, and other mystical traditions. In *The Guru Papers,* published in 1993, they granted that mystical visions can be transformative in the best sense; they can "alter one's relationship to daily life and also profoundly change the way one approaches death and dying." The trouble begins when we translate our visions into ideologies, such as the oneness doctrine.

The oneness doctrine appeals to many modern Westerners, Alstad and Kramer suggested, because it seems less authoritarian and more abstract — and hence easier to reconcile with science — than patriarchal Western theologies. By exalting self-abnegation and renunciation of the world as supreme virtues, Hinduism and Buddhism also appear to offer an antidote to our innate selfishness. But mystical oneness is riddled with con-

---

*The religious scholar Robert Forman told me that awe was not a vital component of his mystical awareness. His first mystical experiences occurred in the early 1970s, shortly after he began practicing Transcendental Meditation. In 1995, he achieved what he described as a "permanent shift in vision," which seemed to correspond to what Ken Wilber calls permanent nondual awareness — that is, enlightenment. Forman described enlightenment as a preternatural calm or equilibrium that comes from recognizing the eternal, universal stillness at the core of your finite, ephemeral self. He felt amazement early on in his spiritual journey, when his mystical perceptions were novel, but over the years his wonder faded. "The amazement came for me when it was fresh and new. But then it becomes steady, and it's not amazing anymore. It just *is.*" Forman described his mystical outlook in *Mysticism, Mind, Consciousness* (State University of New York Press, Albany, 1999).

tradictions, Alstad and Kramer declared: It takes an individual, after all, to experience oneness; moreover, oneness "has within it a hidden duality" that leads to hierarchical social divisions.

Buddhism and Hinduism postulate the existence of certain rare beings who have transcended their individuality and thus experience oneness in a profound and abiding fashion. These supposedly enlightened gurus and avatars often insist that others will achieve enlightenment only through total surrender to them. "The very nature of any structure that makes one person different [from] and superior to others . . . breeds authoritarianism," Alstad and Kramer stated. Hindu ashrams, Buddhist monasteries, and other organizations founded on the oneness principle are usually authoritarian — and patriarchal. It is no accident, they said, that the oneness doctrine is an export of India, which historically has been a highly stratified culture.

Seen through the lens of *The Guru Papers*, some mystical rhetoric appears not paradoxical but Orwellian: Only through submission will we find true liberation. All are one, but some are more one than others. And if spirituality is defined as that which helps us to embrace life in all its painful glory, then enlightenment — which goes hand in hand with oneness — is an anti-spiritual concept. It suggests that life is a problem that can be solved, a riddle that can be answered, a cul-de-sac that can be, and should be, escaped.

Huston Smith contends that enlightenment — defined as supernatural wisdom, grace, moral perfection — is best viewed as an ideal unattainable by mere mortals. Smith shows his own down-to-earth wisdom with this assertion. The notion that we mere mortals can achieve some sort of transcendent, divine state is understandably alluring. But if we judge it by its fruit, as William James recommends, it does not fare well. Belief in the myth of total enlightenment too often turns spiritual teachers into narcissistic monsters, and their students into mindless slaves.

Some bad gurus may be sociopaths without a mystical bone in their bodies, but mystical experiences may exacerbate or instigate sociopathic behavior. Like an astronaut gazing at the earth through the window of his spacecraft, the mystic sees our existence against the backdrop of infinity and eternity. This perspective may not translate into compassion and empathy for others. Far from it. To the mystic, human suffering and death may appear laughably trivial.

The danger posed by a mystical sociopath is compounded if he and his devotees believe that as an enlightened being he has transcended the morality that binds us ordinary mortals. Someone as sensible as James Austin has propagated this insidious notion. In *Zen and the Brain* he declared that "'wrong' actions won't arise when a brain *continues* truly to express the self-nature intrinsic to its kensho experiences." Better to recognize that even the most apparently enlightened masters "have feet of clay — all, no exceptions," as Ken Wilber told me.* Mystical achievement, I suspect, is as closely correlated with moral maturity as scientific or artistic achievement is. In other words, it is not closely correlated at all.

I admit that, as a result of my entheogenic experience in 1981, I still find oneness metaphysically creepy. I keep thinking about the nirvana paradox: If nirvana is so great, why does God create? James Austin's first koan asked: "When all things return to the one, where is the one returned to?" Good question. The reduction of all things to one thing is arguably a route to oblivion; one thing equals nothing. I am hardly alone in fretting over this dilemma. The Hindu sage Ramakrishna no doubt had it in mind when he said, "I want to taste sugar; I don't want to be sugar."

Isn't some separation from ultimate reality necessary for us to appreciate it? Do we really want to live in a world in which there is no other, in which there are no selves but only a single Self? Is that heaven or a solipsistic hell? The Victorian poet G. K. Chesterton raised these questions in his poem "The Mirror of Madmen." The poem's narrator dreams that he has ascended to heaven, where he finds to his horror that all of its denizens — other ascended souls, saints, and angels — have the same face, his face. To his immense relief he awakens just before he sees the face of God.

The movie *Being John Malkovich* presents an areligious version of this nightmare. In the film, a puppeteer, played by John Cusack, discovers that an air-conditioning shaft in the office where he works serves as a portal into the brain of the actor John Malkovich. Those who enter Malkovich's mind see what he sees and feel what he feels. Malkovich, playing himself, even-

---

*A more radical proposal, mentioned by Wilber in *Grace and Grit* (Shambhala, Boston, 1991, pp. 370–373), is that we are *all* enlightened. That is the message of an esoteric Tibetan Buddhist doctrine known as Dzogchen, which Wilber began studying in the 1980s. As he explained it, Dzogchen teaches that there "is nothing you can try to do, or try not to do, to get basic awareness, because it already and fully is" (p. 371). Wilber tied himself in knots trying to make the Dzogchen doctrine compatible with his conviction that enlightenment can be achieved only after years of arduous meditative training, if then.

tually learns about the portal and enters it. He finds himself in a restaurant in which everyone — employees and customers, men and women, even a little girl — has his face. They are all chattering blandly at each other, saying, "Malkovich, Malkovich, Malkovich, Malkovich." Thou Art Malkovich.

As Stanislav Grof told me, various theological doctrines suggest that God cannot bear to stay forever in a state of absolute oneness; that is why He created this flawed, fractured world. That is the implication of the Kabalist doctrine of *tsimtsum*, which holds that creation occurs when God somehow withdraws from Himself; of Meister Eckhart's cryptic comment that "God waxes and wanes"; and of Ken Wilber's remark that not even God likes to eat dinner alone. So if enlightenment is defined as a state of perfect, permanent unity with all things, then not even God is, or can be, totally enlightened.

The ultimate fantasy of some mystical enthusiasts is that one day humanity will be transformed into an "Omega race of humane beings, *all* Buddha-like in their degree of enlightenment," as Austin put it. Plausibility aside, is this destiny desirable? Do we really want to live in the Buddhaverse? The question can be put another way: What if neurotheologians someday discover a mystical technology — a superpsychedelic, a God machine that whispers to our brain cells in their own language, a genetic switch that boosts our brains' production of endogenous DMT — that can reliably induce blissful mystical experiences with no pathological side effects? What would be the consequences for civilization?*

One mystical expert who has addressed this issue is the psychedelic chemist Sasha Shulgin. According to the system he devised for rating the effects of different compounds, the ultimate altered state is a plus four. Shulgin has defined a plus-four experience as "a rare and precious transcendental state," a blissful "peak experience," in which one feels "a connectedness with both the interior and exterior universes."

Shulgin added: "If a drug (or technique or process) were ever to be

---

*This endogenous-DMT proposal came from the psychiatrist and DMT investigator Rick Strassman. In July 2001 I asked Strassman by e-mail if he thought scientists would ever produce a mystical technology so potent that everyone in the world would become mystically awakened. Strassman replied: "I can envision a situation where a cold virus is tinkered with to turn on our methylating enzymes [thereby boosting our endogenous-DMT production], spreads around the world in a couple of years, and there you have it. Not such a weird idea, really. I think, though, there would still be lots of room for creativity and evolution, but just not the way we're used to thinking about it."

discovered which would consistently produce a plus-four experience in all human beings, it is conceivable that it would signal the ultimate evolution, and perhaps the end of the human experiment." I was curious about the somewhat ominous tone of this passage, given that it was written by a man who has dedicated his life to finding new-and-improved entheogens. I e-mailed Shulgin two questions: Does he believe that science will ever produce a drug or other mystical technology that reliably produces plus-four experiences? And what would be the consequences for humanity of such a discovery? Shulgin replied:

> Could such a drug or process someday be discovered? Certainly. Creativity is a continuing process within the human animal, and every new achievement is, in time, surpassed by an even more remarkable one. This is true in inventions, in the miniaturization and acceleration of processes, and certainly in the creation of new drugs. But consider the consequences of the discovery of a drug that produced a continuing plus-four state. Perpetual bliss. In eternal life with no negatives, no anger, no sadness, no competition. There would also be no motivation, no urge to change anything and, I am afraid, no creativity. The evolutional pressure for survival no longer exists. I believe that this would be the end of our species.

In other words, the perfect mystical technology could bring about the end of novelty — or at least human-generated novelty.

Free will and other consolations

I can let go of the oneness meme, the enlightenment meme, and the Buddhaverse meme, but there is one meme that I still need: the free will meme. Susan Blackmore contends that free will is an illusion, and I can understand why. Each moment of our lives is the outcome of a vast web of causes and effects leading back to the big bang itself, the creation event that supposedly set everything in motion. We didn't ask for any of this, so how free can we be? I can see the potential benefits of disbelief in free will, too. In principle, if you expend less energy regretting past decisions and fretting over future ones, you should be better equipped to appreciate the vital present.

But free will *must* exist if some creatures have more of it than others, in the same way that truth must exist if some statements are truer than others. My daughter and son have more free will — more choices to consider and select from — than they did when they were infants. I have more free will than my children do. I also have more than adults my age suffering from schizophrenia or obsessive-compulsive disorder.

Moreover, our belief in free will has social value. It provides us with the metaphysical justification for ethics and morality. It forces us to take responsibility for ourselves rather than entrusting our fate to Jehovah or Allah or the Tao or the timewave. We must accept that things will get better and better only as a result of our efforts, not because we are fulfilling some preordained supernatural plan. If free will is an illusion, it is one that we need — that I need — even more than God. I have no choice but to choose free will.

Of course, I can't be sure that free will exists. There is very little I am sure of, actually. I have ended this mystical inquest as I began it — mystified, or "convictionally impaired," to borrow a phrase from Huston Smith. I believe — I *know* — that our existence is infinitely improbable, a miracle transcending any possible explanation. But I have no idea whether this miracle is eternal or ephemeral. My view of mysticism dispenses with most of the other traditional consolations of spirituality. It does not promise us rebirth, or paranormal powers, or saintly moral wisdom, or immunity from suffering, or heaven or a Buddhaverse or meaning or justice beyond what we create for ourselves. It does not give us the starring role in some divine drama. But it leaves us with one kind of consolation that is enough for me.

Epilogue

Winter Solstice

After I returned home from my ayahuasca session, Suzie pretended that she was feeling better, but I knew she was still grieving for Lena. Her melancholy was infectious, and for several weeks I, too, slipped into a dark mood. I kept waking up in the middle of the night panic-stricken, overwhelmed by the certainty of Suzie's death, my children's death, everyone's death. I remembered something that Ann Shulgin had told me, that when we really *see*, we stand on a knife edge between fear and exhilaration — "because both exist: that terrible darkness and that absolute life." I felt as though I could see only the darkness. Some spiritual teachers advise that brooding over death can help us appreciate life. But why seek death out? It will find us sooner or later. The best spiritual advice is the simplest: Pay attention. See! Or rather, cherish. Cherish what you have before it's gone.

I did my best to console Suzie, and she me, and at some point it occurred to me that this is the only consolation that matters. I remembered something Huston Smith had said during my first conversation with him, when I asked him how he reconciled his faith in the loving, just God of Christianity with the injustice of human suffering. At first Smith gave me a complicated answer about how the infinite must entail all possibilities, including evil. But then, as if acknowledging the fatuity of all theodicies, Smith told me the story about the minister who comforted a woman in despair simply by clasping her hand and weeping with her. Does God care? Who knows? But we care. Our solace — and salvation, if we can be saved — will come not from God's compassion but from ours.

Ken Wilber made the same point implicitly in *Grace and Grit,* which told the story of how he and his wife Treya coped with her breast cancer. One night while they were in Germany, where Treya was receiving a last-ditch treatment, Wilber was overcome with despair. "My whole fucking life is a shambles," he thought. "I've given it all up for Treya, and now Treya, I'll kill her, is going to die." He wandered into a pub and started drinking. Sensing his distress, a group of old German men pulled him off his bar stool and made him dance a polka with them. Alternately crying and laughing as he danced, Wilber felt his self-pity washing out of him. "I would like to claim that my big satori . . . came from some powerful meditation session with blazing white light," Wilber wrote. "But it happened in a little pub with a bunch of kindly old men whose names I do not know and whose language I did not speak."

In retrospect, I intuited this lesson on the morning after my ayahuasca trip. When it was my turn to untie the knot I had made in the prayer rope, I told my companions that the best part of the session had been getting to know all of them. The words came out sounding so mawkish that at the time I immediately regretted them; I doubted my own sincerity. Only later did I realize how much I meant what I said. Immersed in the bizarre DMT visions, I had felt utterly alone. The universe seemed cold and inhuman, and I was on the outside looking in. But those hallucinations were interrupted by moments of communion with my comrades. I felt that sense of common humanity when Tony asked me if I was okay when hallucinations were overwhelming me; when Linda opened the front gate of the ranch for me; when I was jawing with Tony, Kevin, Blaed, and Michael on the ridge overlooking the ocean. I was looking for consolation in the stars, in visions, in mystical gnosis, but the only consolation I found that night was human companionship.

I should have learned this lesson long ago. In the aftermath of my drug trip in 1981, the blackness I had seen at the heart of things seemed more real than the world in which I lived. Not until I met and fell in love with Suzie almost a year after the trip did my estrangement from life and from my own self finally subside. Mysticism did not save me; it was that from which I needed to be saved.

Of course love, which most consoles us, also most desolates us. My wife's heart was broken by a crow. A crow! Buddha advised us to avoid heartbreak by loving less — or rather, by substituting an ethereal, cosmic, divine Love for our painful, mortal, human love. But maybe Buddha didn't have much of a heart himself. After all, he began his quest for enlightenment by leaving his wife and child. The price we pay for having hearts is that they can be broken. Mystical theologies too often console us by devaluing our mortal lives. They tell us that we and those we love are as ephemeral as dreams, compared to the eternal, infinite nothing from which we came and to which we will return. If this is a dream, I don't want to awaken.

As summer gave way to fall, Suzie regained her witchy equanimity. She drew strength from nature, as always. She took long walks in the woods by our house, keeping an eye out for her favorites, goshawks, pileated woodpeckers, and cardinals. She befriended a trio of crows living nearby. She put chicken necks, leftover spaghetti, and other kitchen scraps on a flat rock in the woods. When she whistled, the three crows careened down from the trees, cawing raucously. It was almost like old times.

Suzie grew excited as December 22 approached. This would be the first winter solstice in 133 years, she told me, when the moon would be full. Also, the earth would be almost as close as it gets to the sun, and the moon would be at its near point to the earth — 222,000 miles away compared to its average distance of 239,000 miles. All this meant that the moon would be preternaturally bright on the evening of December 22 — bright enough to drive without headlights, some were saying on the Internet. Suzie decided that our family should celebrate this celestial event in a field beside our house. I would build a fire at a spot that she'd prepare in advance. When the fire got going, she and the kids, Mac and Skye, would join me.

On the night of the solstice, after we ate an early dinner, I pulled on a coat and gloves and headed outside with a candle lantern, matches, and a stick of paraffin-soaked kindling. The moon, already ascended above the trees, was indeed brilliant. Gazing at it, or even at its star-dimming penumbra, hurt my eyes. The shadows cast by the skeletal cherry tree in our front yard and by my body looked sharper-edged than on a summer day.

I trudged down our driveway and passed beneath two arching dogwoods into the field beyond. Following Suzie's directions, I found a circle of stones she had assembled on the field's far border. In the center of the circle was a heap of sticks that Suzie, Mac, and Skye had gathered earlier that day. Putting my lantern down, I thrust the chunk of kindling into the pile of sticks and put a match to it. Before long, twigs were flaring and snapping.

I stretched out on the grass and admired my handiwork. A few minutes later I heard voices and saw three yellow lights bobbing at the edge of the field. When Mac and Skye spotted the fire, they called out and ran toward me, their lanterns swinging perilously. Suzie strolled behind them, holding her lantern in one hand and a pail of water in the other.

We hunched close to the fire, encircled by the stones. Suzie and I took turns telling tales about the Man in the Moon, Orion, and other sky dwellers — or we tried to, anyway. Mac kept interrupting our narratives to ask for clarification. Skye was less interested in our stories than in the popping, incandescent fire. She pulled a stick out and inspected the burning end so closely that I warned her she was about to set her hair on fire. After the stick flickered out, Skye shoved it so far back into the fire that her mitten smoked. I nagged her about that, too, but she ignored me. Mac, seeing the fun Skye was having, joined her in poking and prodding the fire.

It occurred to me that I should stop worrying about the kids and appreciate the moment. Pay attention. *See!* A decades-old memory popped into my head: I was climbing a mountain on the coast of Maine on a late-summer day and was looking east toward the Atlantic when I noticed a gibbous moon. And suddenly I *saw* it, not the old familiar moon, as blandly symbolic as a poker chip, but the *moon,* a gigantic ball of pocked rock hovering miraculously in a blue abyss.

Inspired by this memory, I squinted at our wintry moon — the same moon! — determined to *see* it. In vain. It was still too dazzling, and thoughts — of that afternoon in Maine long ago, of all the reasons why the moon was so brilliant tonight — drifted like my own misty exhalations across my line of sight. The last straw was my recollection that the moon is the Zen emblem of enlightenment. No chance for seeing it now.

My gaze fell back down to earth, to our stone-enclosed circle, bobbing like a dinghy on the silvery tufts and hillocks of the field. Mac and Skye, their eyes and skin aglow with marmalade-colored light, were still tormenting the fire with sticks. Suzie, clasping her knees, was staring up at the starry northern sky. I wanted to ask her if she could see something up there, something that had eluded me, but I left her to her thoughts.

The cold seeped through our clothes. Time to head back, Suzie said. Mac and I took turns pouring water from the pail onto the fire. After the last ember went dark, I led the way from the circle of stones toward the lights of our house, barely visible through a stand of ghostly trees. Mac and Skye, somewhere behind me, began chattering about diamonds. Where did the diamonds come from? Skye asked. That's frost, Suzie replied. What's frost? Mac asked. Frost is little pieces of ice, Suzie explained, like tiny ice cubes. When the moonlight goes through them, they sparkle like diamonds. Ahhh, said the kids. Mystified, I stared at the pallid earth, and all of a sudden I saw them, diamonds, sapphires, emeralds, rubies, points of pure iridescence, sprinkled like fairy dust on the path leading back to our home.

Acknowledgments

In the course of writing this book, I have been aided by more people — including strangers encountered at parties or elsewhere who offered me theological suppositions, tall psychedelic tales, book recommendations, and so on — than I can mention or recall here. Those who gave me valuable feedback on parts or all of the book include Diana Alstad, Chris Bremser, Michael Carroll, Tucker Clark, Charles Cook, Martin Goodman, Marcy Gordon, John Halpern, Don Hill, Ruth and Robert Hutchinson, Joel Kramer, Luis Eduardo Luna, Dennis McKenna, David Rothenberg, Ann and Alexander Shulgin, Huston Smith, Gary Stix, James Thornton, Franz Vollenweider, Karen Wright, and Robert Wright.

Others who helped me in one way or another include Jensine Andreson, Leonard Angel, James Austin, Sarah Banker, Stephen Batchelor, Susan Blackmore, Andrew Cohen, Jack Cowan, John Cronin, Wade Davis, Adolf Dittrich, Rick Doblin, Robert Forman, Robert Frager, Frank Geer, Mark Geyer, Neal Goldsmith, Andrew Greeley, Arthur Green, Alex Grey, Lester Grinspoon, Stanislav Grof, Stewart Guthrie, J. P. Harpignies, Jennine Heller, Albert Hofmann, Adrian Hofstetter, Jean Houston, Robert Jesse, Steven Katz, Jay Kinney, Paul Krassner, Evgeny Krupitsky, Chisho Maas, Bernard McGinn, Terence McKenna, Ralph Metzner, Andrew Newberg, David Nichols, Jonathan Ott, Michael Persinger, V. S. Ramachandran, Christian Ratsch, Tom Roberts, Christian Sharffetter, Roger Shattuck, Michael Shermer, Ronald Siegel, Richard Smoley, Alan Stein, David Steindl-Rast, Rick Strassman, Rick Tarnas, Charles Tart, Robert Thurman, Francisco Varela, Francis Vaughan, Roger Walsh, and Ken Wilber.

I am grateful to my agents, John Brockman and Katinka Matson, for helping me shape my proposal and finding a taker for it, to Laura Van Dam for bringing me to Houghton Mifflin and giving me encouragement early on, to Eamon Dolan for helping me turn a mere manuscript into a book, and to Larry Cooper for his sharp-eyed final edit. Special thanks to Suzie for helping me in every possible way, and then some.

Notes

Introduction: Lena's Feather

PAGE

2 "vine of the dead": I found this translation of *ayahuasca* in *Psychedelic Shamanism,* by Jim DeKorne, Loompanics Unlimited, Port Townsend, Wash., 1994, p. 91.

4 "the mind of God": *A Brief History of Time,* by Stephen Hawking, Bantam Books, New York, 1988, p. 175.

Charles Tart . . . complained: *Altered States of Consciousness,* third edition, edited by Charles Tart, HarperCollins, New York, 1990, p. 12.

6 begins in mist: Jean Houston, personal communication, August 5, 1999.

mysticism meets . . . four criteria: *The Varieties of Religious Experience,* by William James, Macmillan, New York, 1961, pp. 299–300.

"immense joyousness": *Cosmic Consciousness,* by Richard Bucke, Causeway Books, New York, 1974, p. 8.

7 "The mystic vision is not a feeling": *Forgotten Truth,* by Huston Smith, HarperSanFrancisco, 1992, p. 112.

"the Way Things Are": I found this quote in *Zen and the Brain,* by James Austin, MIT Press, Cambridge, 1998, p. 19. Austin cited Greeley's book *Ecstasy: A Way of Knowing,* Prentice-Hall, Englewood Cliffs, N.J., 1974.

A survey carried out in the 1970s: *The Sociology of the Paranormal: A Reconnaissance,* by Andrew Greeley and William McCready, Sage Research Paper, vol. 3, series 90-023, Beverly Hills, Calif., 1975. Greeley and McCready also reported their results in "Are We a Nation of Mystics?," *New York Times Magazine,* January 26, 1975, p. 6. I found Greeley's work summarized in Austin, *Zen and the Brain.* Austin (p. 454) said that the major triggers of mystical experiences, as reported by Greeley's subjects, were music (49 percent), prayer (48 percent), the beauties of nature (45 percent), quiet reflection (42 percent), childbirth (20 percent), and lovemaking (18 percent).

A British poll determined: This poll, carried out by Alister Hardy and described in his book *The Spiritual Nature of Man* (Oxford University Press, Oxford, 1979), was also cited in Austin, *Zen and the Brain,* p. 20.

PAGE

8 *Mysticism* derives: *Forgotten Truth*, by Huston Smith, HarperSanFrancisco, 1992, p. 110. For further discussion of the etymology of *mysticism*, see also *Mysticism*, by F. C. Happold, Penguin Books, New York, 1961, p. 18.

10 "the attainment of a state of spiritual truth": *Intelligence Reframed*, by Howard Gardner, Basic Books, New York, 1999, p. 56.

11 "Psychedelic drugs are easier to study": *Psychedelic Drugs Reconsidered*, by Lester Grinspoon and James Bakalar, The Lindesmith Center, New York, 1997, p. 239.

you cannot tread the path: *Consilience*, by Edward O. Wilson, Knopf, New York, 1998, p. 262.

13 "heaven, hell, and visions": I interviewed Adolf Dittrich in Basel, Switzerland, on November 14, 1999. He has presented his three-dimensional model in "International Study on Altered States of Consciousness: Summary of the Results," by Adolf Dittrich, S. von Arx, and S. Staub, *German Journal of Psychology*, vol. 9, 1985, pp. 319–339; and in "Psychological Aspects of Altered States of Consciousness of the LSD Type: Measurement of Their Basic Dimensions and Prediction of Individual Differences," *50 Years of LSD: Current Status and Perspectives of Hallucinogens*, edited by A. Pletscher and D. Ladewig, Parthenon, New York, 1994, pp. 101–118.

14 *shruti . . . smriti:* I found these translations of *shruti* and *smriti* in *The Perennial Philosophy*, by Aldous Huxley, Harper & Row, New York, 1944, p. ix.

1. Huston Smith's Perennial Philosophy

15 "Science and Consciousness": The 1999 International Conference on Science and Consciousness convened at the Crowne Plaza Pyramid Hotel in Albuquerque, April 9–14.

Edgar Mitchell: For background on Mitchell and the Institute of Noetic Sciences, see www.noetic.org. The Web site notes that looking at the earth during his 1971 moonwalk mission, Mitchell had a mystical epiphany. "The presence of divinity became almost palpable," he recalled, "and I knew that life in the universe was not just an accident based on random processes . . . The knowledge came to me directly."

16 "as we have discovered": *Dreams of a Final Theory*, by Steven Weinberg, Pantheon, New York, 1992, p. 253.

the ethnopharmacologist Dennis McKenna: Among McKenna's many peer-reviewed publications are "Plant Hallucinogens: Springboards for Psychotherapeutic Drug Discovery," *Behavioural Brain Research*, vol. 73, 1996, pp. 109–115; and "The Scientific Investigation of Ayahuasca: A Review of Past and Current Research," cowritten with J. C. Callaway and Charles Grob, *The Heffter Review of Psychedelic Research*, vol. 1, 1998, pp. 65–77.

17 Leibniz coined the phrase: *The Perennial Philosophy*, by Aldous Huxley, Harper & Row, New York, 1944, p. vii.

PAGE

"human sacrifice and scapegoating": *The World's Religions,* by Huston Smith, HarperSanFrancisco, 1991, p. 4.

"a stained-glass window": Ibid., p. 386.

19 As we sat down that evening: I interviewed Smith in Albuquerque on April 10, 1999.

21 "strange, weird, uncanny": These descriptions of Smith's first entheogenic experience can be found in his book *Cleansing the Doors of Perception,* Jeremy Tarcher/Putnam, New York, 2000, pp. 9–13.

22 Smith epitomizes "mature spirituality": *Shadows of the Sacred,* by Frances Vaughan, Quest Books, Wheaton, Ill., 1995, p. 250.

Jean Houston described him: Houston discussed Smith when I interviewed her at her home on September 2, 1999.

religions " 'flesh out' God's nature": "Is There a Perennial Philosophy?," by Huston Smith, *Journal of the American Academy of Religion,* Fall 1987, p. 562.

23 "fractures into multiple forms": Smith, *Cleansing the Doors of Perception,* p. 11.

24 "The Western hunt for knowledge": *Forgotten Truth,* by Huston Smith, HarperSanFrancisco, 1992, p. 126.

"never had any trouble with the scientists": In ibid., p. 11, Smith recalled a scientist at MIT who summed up the difference between his work as a scientist and Smith's as a representative of the humanities with a devastating double entendre: "I count and you don't."

25 Huxley saw "what Adam had seen": *The Doors of Perception,* by Aldous Huxley, Harper & Row, New York, 1990, p. 17.

"I am not so foolish": Ibid., p. 73.

27 The findings reported by Pahnke: Pahnke and a coauthor, William Richards, described the Good Friday experiment in "Implications of LSD and Experimental Mysticism," in *Altered States of Consciousness,* third edition, edited by Charles Tart, HarperCollins, New York, 1990, pp. 481–515. For a contrary view of drug-induced versus spontaneous mystical experiences, see "Cosmic Consciousness and Psychedelic Experiences: A First-Person Comparison," by Allan Smith and Charles Tart, *Journal of Consciousness Studies,* vol. 5, no. 1, 1998, pp. 97–107.

One participant who remembers: Huston Smith recalled the Good Friday experiment in *Cleansing the Doors of Perception,* p. 99–105.

28 A recent review of the Good Friday experiment: "The Good Friday Experiment: A Long-Term Followup and Methodological Critique," by Rick Doblin, *Journal of Transpersonal Psychology,* vol. 23, no. 1, 1991, pp. 1–28.

29 A Talmudic legend: I found this story in *A History of God,* by Karen Armstrong, Ballantine Books, New York, 1993, pp. 212–213.

30 William James, after inhaling nitrous oxide: The passage to which Huston Smith is referring can be found in *The Varieties of Religious Experience,* by William James, Macmillan, New York, 1961, p. 306. James wrote of his nitrous oxide experience: "It is as if the opposites of the world, whose contradictoriness and

PAGE

conflict make all our difficulties and troubles, were melted into unity. Not only do they, as contrasted species, belong to one and the same genus, but *one of the species,* the nobler and better one, *is itself the genus, and so soaks up and absorbs its opposite into itself*" (italics in the original).

31 a "black-haired youth with greenish skin": Ibid., p. 138. For a fascinating discussion of this episode, see "William James and the Case of the Epileptic Patient," by Louis Menand, *New York Review of Books,* December 17, 1998, pp. 81–93. Menand argued persuasively that James's distress was caused by his recognition that, as Menand put it, "moral worth does not protect us against disaster"; suffering is meted out randomly, not according to any system of justice.

32 But all evil is arguably natural evil: The British theologian Paul Fiddes made this argument in *The Creative Suffering of God,* Oxford University Press, Oxford, 1992, pp. 33–34.

"cosmic paranoia" or "demonization": *The Foundations of Mysticism,* by Bernard McGinn, Crossroad, New York, 1991, p. 19. For a more positive view of gnosticism, see *The Gnostic Gospels,* by Elaine Pagels, Vintage Books, New York, 1979. Pagels, a prominent religious scholar, depicted gnosticism as a kind of ur–New Age religion that was anti-authoritarian, anti-hierarchical, pro-female, and focused on subjective experience rather than moralistic behavior as the basis of spirituality. Pagels seemed almost to regret that gnosticism had been stamped out by orthodox Christianity. The version of Christianity that has been passed down to us, she suggested, survived less because of its superior virtues than because it was more suited to institutionalization and hence propagation. But Pagels acknowledged that gnosticism had its faults: It exhibited tendencies toward elitism, esotericism, and contempt for the world of flesh. Moreover, its insistence that only our subjective inner experience matters, and not our behavior, verged on solipsism.

"All Right": I found this quote from Aldous Huxley in *One Taste,* by Ken Wilber, Shambhala Publications, Boston, 1999, p. 243.

34 multiple-personality disorder was all the rage: For an account of how modern psychiatry became enamored of multiple-personality disorder, see "Sybil: The Making of a Disease," by Mikkel Borch-Jacobsen, *New York Review of Books,* April 24, 1997, pp. 60–64.

2. Attack of the Postmodernists

36 a two-day conference: "Mystics: A Conference on Presence and Aporia" was held May 13–14, 1999, at the University of Chicago.

37 "Academic religion is the killing jar of Spirit": *One Taste,* by Ken Wilber, Shambhala Publications, Boston, 1999, p. 312.

38 "The more it climbs": *The Foundations of Mysticism,* by Bernard McGinn, Crossroad, New York, 1991, p. 159.

PAGE

you can't "whistle" about it either: After the Chicago meeting I learned that when Kevin Hart delivered this "whistle" quip he was quoting the scholar Frank Ramsey (see *Forgotten Truth,* by Huston Smith, HarperSanFrancisco, 1992, p. 58).

40 God "becomes a possibility": McGinn, *The Foundations of Mysticism,* p. xviii.

41 "strong biases": *Mysticism and Philosophical Analysis,* edited by Steven Katz, Oxford University Press, New York, 1978, p. 2.

"The cure for": Ibid., p. 1.

42 "reflect an underlying similarity": Ibid., p. 23.

"There is no *philosophia perennis*": Ibid., p. 24.

"There are NO pure": Ibid., p. 26.

43 I met Katz: I interviewed Steven Katz at Boston University on October 5, 1999.

48 "there is no mysticism as such": *Major Trends in Jewish Mysticism,* by Gershom Scholem, Schocken Books, New York, 1946, p. 6. See also Scholem's discussion of the Kabalist doctrine of *tsimtsum* (pp. 260–261) and of the importance of sexuality and marriage for those on the mystical path (p. 235).

"cultural solipsism": "Is There a Perennial Philosophy?," by Huston Smith, *Journal of the American Academy of Religion,* vol. 55, no. 3, 1987, p. 560. A rebuttal from Katz and a counterrebuttal from Smith can be found in the *Journal of the American Academy of Religion,* vol. 56, no. 4, 1988, pp. 751–759.

49 "the scandal of particularity": *The World's Religions,* by Huston Smith, HarperSanFrancisco, 1991, p. 308.

50 "Any God who": *The Varieties of Religious Experience,* by William James, Macmillan, New York, pp. 276–277.

"principles of impartiality and fair play": Ibid., p. 307.

51 Council on Spiritual Practices: See the council's Web site at www.csp.org.

52 "mere regressions or repressions of sexuality": McGinn, *The Foundations of Mysticism,* p. 343.

transpersonal psychology: For an overview of transpersonal psychology, see *Paths Beyond Ego,* edited by Roger Walsh and Frances Vaughan, Jeremy Tarcher/Putnam, New York, 1993. Walsh, a psychiatrist, and Vaughan, a psychologist and Walsh's wife, are important figures in transpersonal psychology. The major transpersonal organization is the Association of Transpersonal Psychology, in San Francisco (www.atpweb.org).

Gurdjieff proclaimed that the moon: The bizarre beliefs and behavior of Gurdjieff, Rajneesh, and other gurus were discussed by the British psychiatrist Anthony Storr in *Feet of Clay,* Free Press, New York, 1996. Trying to distinguish guru-led cults from religions, Storr made this sardonic observation: "Idiosyncratic belief systems which are shared by only a few adherents are likely to be regarded as delusional. Belief systems which may be just as irrational but which are shared by millions are called world religions" (p. 203). Storr included Jung and Freud in his list of gurus.

PAGE

52 Meher Baba . . . Nityananda . . . Rajneesh . . . Trungpa . . . Osel Tendzin: My descriptions of these gurus are based on various published sources, including Storr, *Feet of Clay; Holy Madness,* by Georg Feuerstein, Paragon House, New York, 1991; *The Book of Enlightened Masters,* by Andrew Rawlinson, Open Court, Chicago, 1997; "Crimes of the Soul," by Jill Newmark, Marian Jones, and Dennis Gersten, *Psychology Today,* March/April 1998, p. 58; and "Vajra Regent Osel Tendzin, 47, Former Leader of Buddhist Sect" (obituary), *Boston Globe,* August 28, 1990, p. 35. For a gripping account of Shoko Asahara and the Aum Shinrikyo cult, see *Destroying the World to Save It,* by Robert J. Lifton, Henry Holt, New York, 2000. For a thoughtful overview of gurus and cults by a psychiatrist, see *Cults,* by Marc Galanter, Oxford University Press, New York, 1999.

a true Christian must "become a fool": Feuerstein, *Holy Madness,* p. 9.

"the path of blame" . . . Malamatis: Ibid., p. 18.

54 "We cannot know the inner man": Ibid., p. 139.

3. The Weightlifting Bodhisattva

55 *New York Times* revealed: "Erudite and Groovy," by James Atlas, *New York Times,* August 7, 2000, p. A19.

a letter from Chopra: *New York Times,* August 17, 2000, p. A28.

56 "the Enlightenment of the East": *The Marriage of Sense and Soul,* by Ken Wilber, Broadway Books, New York, 1998, p. 211.

57 "a holarchy of extended love": *Integral Psychology,* Shambhala Publications, Boston, 2000, p. 189.

riddled with pitfalls: Wilber discussed "psychic inflation" and other pathologies that can arise on the path to enlightenment in "The Spectrum of Pathologies," in *Paths Beyond Ego,* edited by Roger Walsh and Frances Vaughan, Jeremy Tarcher/Putnam, New York, 1993, pp. 156–159.

58 "I simply reverted to what I am": *One Taste,* by Ken Wilber, Shambhala Publications, Boston, 1999, p. 69. In December 2001, Wilber informed me in an e-mail message that this episode occurred in 1995.

"Yesterday I sat": Wilber, *One Taste,* pp. 82–83.

59 Wilber agreed to meet me: I interviewed Ken Wilber on October 20, 1999. For an affectionate portrait of Wilber (more affectionate than mine, at any rate), see *What Really Matters,* by Tony Schwartz, Bantam Books, New York, 1995, pp. 339–374.

a Swiss chalet and a Japanese bathhouse: "Up Close and Transpersonal with Ken Wilber," by Mark Matousek, *Utne Reader,* July/August 1998, p. 53.

62 the self-proclaimed avatar Da Free John: My description of Da Free John is primarily based on a book by a former devotee, *Holy Madness,* by Georg Feuerstein, Paragon House, New York, 1991, pp. 80–100; as well as *The Book of Enlightened Masters,* by Andrew Rawlinson, Open Court, Chicago, 1997, pp. 221–230; and "Scandalous Times of Perfect Masters," by Don Lattin, *San Francisco Chronicle,* August 3, 1988, p. B3.

PAGE

"a Spiritual Master and religious genius": I found this quotation from Wilber in Rawlinson, *The Book of Enlightened Masters,* p. 229.

63 life is getting better and better: For a compelling argument by a first-rate science journalist that things are, or could be, getting better and better as a result of natural evolutionary processes that promote symbiosis and cooperation — and perhaps because of divine influence — see *Nonzero,* by Robert Wright, Pantheon, New York, 2000.

64 he struck her: *Grace and Grit,* by Ken Wilber, Shambhala Publications, Boston, 1991, p. 154.

favorably reviewed by: This review, of Wilber's book *The Marriage of Sense and Soul,* was "Can There Be a Science of Spirituality?," by Jeff Minerd, *Skeptical Inquirer,* January/February 2000, pp. 53–54.

65 fawning letters from admirers: See, for example, Wilber, *One Taste,* pp. 45–53.

"I-I": In ibid., p. 59, Wilber attributed the term "I-I" to the Hindu sage Ramana Maharshi.

"If Spirit does exist": Wilber, *The Marriage of Sense and Soul,* p. xi.

"not 'omniscience' but 'ascience'": Wilber, *One Taste,* p. 152.

67 "mostly of the male sex": *Cosmic Consciousness,* by Richard Bucke, Causeway Books, New York, 1974, p. 55.

"Aryan" . . . "the negro race": Ibid., p. 49.

68 "This new race": Ibid., p. 318.

"the vague feeling": Wilber, *One Taste,* p. 209.

"It's no fun . . . *can* you do": Ibid., p. 207.

70 "God is crazy just like me": I found this quote from Da Free John in Feuerstein, *Holy Madness,* p. 97.

71 "radical transformative spirituality is extremely rare": "A Spirituality that Transforms," by Ken Wilber, *What Is Enlightenment?,* Fall/Winter 1997, p. 26.

Wilber was the subject of an experiment: Wilber discussed his EEG readings in *One Taste,* pp. 75–77, 277–278.

4. Can Neurotheology Save Us?

73 "Although the field is brand new": "God and the Brain," by Sharon Begley, *Newsweek,* May 7, 2001, p. 53.

74 "seems to be identical": *The Mystical Mind,* by Andrew Newberg and Eugene D'Aquili, Fortress Press, Minneapolis, 1999, p. 200.

75 skeptics "can no longer dismiss": Ibid., p. 206.

"not on delusional *ideas*": *Why God Won't Go Away,* by Andrew Newberg, Eugene D'Aquili, and Vince Rause, Ballantine Books, New York, 2001, p. 126. Rause is a journalist.

"It is possible that with the advent": Newberg and D'Aquili, *The Mystical Mind,* p. 207.

76 I met Newberg: I interviewed Andrew Newberg at the Hospital of the University of Pennsylvania on March 5, 2001.

PAGE

78 "higher-than-average levels": Newberg, D'Aquili, and Rause, *Why God Won't Go Away,* p. 108.

79 as many as half of [near-death experiences] are hellish: Newberg and D'Aquili discussed the evidence for "hellish" near-death experiences in *The Mystical Mind,* pp. 125–126.

a sect in the Far East: In ibid., pp. 173–174, Newberg and D'Aquili mentioned a sect whose members reportedly sought states in which "the entire universe is viewed as intrinsically evil and horrible."

81 brain scans of schizophrenics vary: When I asked Newberg later where this research on the neurophysiology of real versus delusional perceptions had been published, he cited three papers: "Where the Imaginal Appears Real: A Positron Emission Tomography Study of Auditory Hallucinations," by H. Szechtman et al., *Proceedings of the National Academy of Sciences,* vol. 95, no. 4, 1998, pp. 1956–1960; "The Functional Anatomy of Auditory Hallucinations in Schizophrenia," by B. R. Lennox et al., *Psychiatry Research,* vol. 100, no. 1, 2000, pp. 13–20; and "Functional Neuroimaging of Hallucinations in Schizophrenia: Toward an Integration of Bottom-Up and Top-Down Approaches," by D. Silbersweig and E. Stern, *Molecular Psychiatry,* vol. 1, no. 5, 1996, pp. 367–375.

82 some of history's greatest religious figures: For a discussion of the pathology of historical religious figures, see "The Neural Substrates of Religious Experience," by Jeffrey Saver and John Rabin, *Journal of Neuropsychiatry,* vol. 9, no. 3, 1997, pp. 498–510.

"mystical by default" . . . "wired": Newberg, D'Aquili, and Rause, *Why God Won't Go Away,* pp. 37, 107.

83 a 1975 survey: Ibid., p. 107. This survey is the same one I cited in the introduction: *The Sociology of the Paranormal: A Reconnaissance,* Sage Research Paper, vol. 3, series 90-023, Beverly Hills, Calif., 1975.

Greeley, who once told me: I spoke to Andrew Greeley by telephone on April 5, 1999.

A British survey: This survey, conducted by Alister Hardy and described in his book *The Spiritual Nature of Man* (Oxford University Press, Oxford, 1979), was cited in Saver and Rabin, "The Neural Substrates of Religious Experience," p. 499.

84 "prodded and poked by": "Meditation Meets Behavioral Medicine," by Jensine Andreson, *Journal of Consciousness Studies,* vol. 7, no. 11–12, 2001, p. 32.

a study . . . by Danish researchers: See ibid., pp. 43–44.

85 "manipulation and control": Ibid., pp. 39–40.

"feedback through the limbic": "The Neurophysiology of Religious and Spiritual Experience," by Andrew Newberg and Eugene D'Aquili, *Journal of Consciousness Studies,* vol. 7, no. 11–12, 2001, p. 261.

Rube Goldberg contraption: I found this analogy in *Mistaken Identity,* by Leslie Brothers, State University of New York Press, Albany, 2001, p. 23. In this book, Brothers, a psychiatrist who has carried out neural research on primates, was

PAGE

critical of attempts of modern philosophers and scientists to explain away the mind-body problem. She also presented stinging critiques of the work of Newberg and D'Aquili, James Austin, and Michael Persinger.

86 Templeton Foundation: For articles about the foundation and the controversy it has provoked in the scientific community, see "Subjecting Belief to the Scientific Method," by Constance Holden, *Science,* May 21, 1999, pp. 1257–1259; and "AAAS Members Fret Over Links with Theological Foundation," by Colin Macilwain, *Nature,* February 24, 2000, p. 819. The foundation's largesse is so vast that it has trickled down to me: Templeton funds helped pay for me to give a lecture on mysticism at the University of San Francisco in March 2000.

The "existence of an absolute higher reality": Newberg, D'Aquili, and Rause, *Why God Won't Go Away,* p. 155.

"Or how do you Mystics": I found this quotation from David Hume in *Faces in the Clouds,* by Stewart Guthrie, Oxford University Press, New York, 1993, p. 177. Guthrie, an anthropologist who argues that religion is a byproduct of our innate anthropomorphism, cited Hume's *Dialogues Concerning Religion.*

87 "evolutionary perspective suggests": Newberg, D'Aquili, and Rause, *Why God Won't Go Away,* p. 126.

a 1979 paper by the Harvard biologists: "The Spandrels of San Marcos and the Panglossian Paradigm," by Stephen Jay Gould and Richard Lewontin, *Proceedings of the Royal Society,* vol. 205, 1979, pp. 581–598.

89 "Every medical student is taught": *Phantoms in the Brain,* by V. S. Ramachandran and Sandra Blakeslee, William Morrow, New York, 1998, p. 175.

90 "But what else *is* there?": Ibid., p. 182.

"the God module": Ibid., p. 175.

5. The God Machine

92 "sensed presence": See "Experimental Induction of the 'Sensed Presence' in Normal Subjects and an Exceptional Subject," by Charles Cook and Michael Persinger, *Perceptual and Motor Skills,* vol. 85, 1997, pp. 683–693. For a good recent summary by Persinger of his research and theories, see also "The UFO Experience: A Normal Correlate of Human Brain Function," in *UFOs and Abductions,* edited by David Jacobs, University Press of Kansas, Lawrence, 2000, pp. 262–302.

as many as 40 percent of his subjects: Persinger cited the 40 percent figure to me when I met him in Sudbury. In "The UFO Experience," he said that anywhere from 27 percent to 80 percent of the subjects of his God machine experiments sensed a presence.

ABC News: ABC News mentioned Persinger's research in a report on science and religion that aired February 23, 1998.

"a radio for speaking to God": "Searching for God in the Machine," by David Noelle, *Free Inquiry,* Summer 1998, p. 54.

PAGE

92 the Canadian magazine *Maclean's:* "The God Machine," by Mark Nichols, *Maclean's,* January 22, 1996, p. 46.

the linkage of mystical states and epileptic seizures: The evidence for this linkage was summarized in "The Neural Substrates of Religious Experience," by Jeffrey Saver and John Rabin, *Journal of Neuropsychiatry,* vol. 9, no. 3, 1997, pp. 498–510.

93 James felt compelled to denounce it: *The Varieties of Religious Experience,* by William James, Macmillan, New York, 1961, p. 303.

Julian Jaynes proposed: *The Origin of Consciousness in the Breakdown of the Bicameral Mind,* by Julian Jaynes, Houghton Mifflin, Boston, 1976.

experiments by . . . Penfield: For lucid discussions of Penfield's experiments, hemispheric studies, and other research related to Michael Persinger's theories, see *The Three-Pound Universe,* by Judith Hooper and Dick Teresi, Jeremy Tarcher, Los Angeles, 1986.

"Is there, one wonders": *The Doors of Perception* (which includes the essay "Heaven and Hell"), by Aldous Huxley, Harper & Row, New York, 1990, pp. 147–148.

Ramachandran mused that his friend: *Phantoms in the Brain,* by V. S. Ramachandran and Sandra Blakeslee, William Morrow, New York, 1998, p. 175.

"nothing but a pack of neurons": *The Astonishing Hypothesis,* by Francis Crick, Charles Scribner's Sons, New York, 1994, p. 3.

"theotoxins": The transpersonal psychiatrist Roger Walsh and the psychologist Frances Vaughan credited Crick with having coined the term *theotoxin* in a book they edited: *Paths Beyond Ego,* Jeremy Tarcher/Putnam, New York, 1993, p. 178.

94 I flew to Toronto: I interviewed Michael Persinger on March 11, 1999.

95 In his books: Persinger's two books are *TM and Cult Mania,* Christopher Publishing, North Quincy, Mass., 1980; and *Neuropsychological Bases of God Beliefs,* Praeger, New York, 1987.

those who respond strongly to meditation: "Transcendental Meditation and General Meditation Are Associated with Enhanced Complex Partial Epileptic-like Signs," by Michael Persinger, *Perceptual and Motor Skills,* vol. 76, no. 1, 1993, pp. 80–82.

98 "Elijah, Jesus, the Virgin Mary": "This Is Your Brain on God," by Jack Hitt, *Wired,* November 2000. I found the article on the *Wired* Web site: www.wired.com/wired/archive/7.11/persinger.html.

A Canadian journalist sensed: The journalist is the radio broadcaster Don Hill, whom I met and interviewed in Sudbury on March 12, 1999. Hill told me that he first saw a ghost in the basement of his home in the early 1990s. A transcript of Hill's session can be found in "Neurobehavioral Effects of Brief Exposures to Weak Intensity, Complex Magnetic Fields within Experimental and Clinical Settings," a paper delivered by Michael Persinger at "Magnetic Fields: Recent Advances in Diagnosis and Therapy," a conference held in London, Ontario, on November 15, 1997.

PAGE

Cotton . . . daydreamed that: "Dr. Persinger's God Machine," by Ian Cotton, *Free Inquiry,* Winter 1996, p. 47. Cotton also described his encounter with Persinger in *The Hallelujah Revolution,* Prometheus Books, Amherst, New York, 1996, p. 197.

99 Persinger "experienced God": Ramachandran and Blakeslee, *Phantoms in the Brain,* p. 175.

temporal-lobe epileptics — who account for more than one third of the . . . epileptics in the U.S.: I found these statistics in "Brain Pacemakers," by Stephen Hall, *Technology Review,* September 2001, p. 42; and "Study Says Surgery Tops Drugs in Epilepsy," by Sandra Blakeslee, *New York Times,* August 2, 2001, p. A12.

activity in the temporal lobes might account for some mystical visions: For a good review of the evidence, see Saver and Rabin, "The Neural Substrates of Religious Experience."

electroshock therapy: I reviewed recent research on shock therapy in *The Undiscovered Mind,* pp. 128–134.

Researchers are now testing whether much milder pulses: For a review of this research, see "Transcranial Magnetic Stimulation and the Human Brain," by Mark Hallet, *Nature,* July 13, 2000, pp. 147–150.

100 "to influence directly the major portion": "On the Possibility of Directly Accessing Every Human Brain by Electromagnetic Induction of Fundamental Algorithms," by Michael Persinger, *Perceptual and Motor Skills,* vol. 80, 1995, p. 797.

the neuroscientist Robert Heath: For accounts of Heath's experiments, see Hooper and Teresi, *The Three-Pound Universe,* 152–161.

101 the neuroscientist José Delgado: See ibid., pp. 145–151. For a discussion relevant to Michael Persinger's suggestion that scientists may someday be able to read our minds remotely, see also the section of Hooper and Teresi's book titled "The Impossibility of Mind Reading," pp. 143–144. Some scientists nonetheless claim they can detect lies by monitoring the brain's EEG signals. See "Truth and Justice, by the Blip of a Brain Wave," by Barnaby Feder, *New York Times,* October 9, 2001, p. F3.

Delgado prophesied: *Physical Control of the Mind,* by José Delgado, Harper & Row, New York, 1969.

Implanted electrical-stimulation devices are now occasionally used: For reviews of this research, see Hall, "Brain Pacemakers"; and "Brain Stimulation: Current Applications and Future Prospects," by Alon Mogilner, Alim-Louis Benabid, and Ali Rezai, *Thalamus and Related Systems,* vol. 1, 2001, pp. 255–267.

102 psi doubters such as Michael Shermer: See Shermer's description of Persinger's work in *How We Believe,* by Michael Shermer, W. H. Freeman, New York, 2000, pp. 66–69. In retrospect, I should have been tipped off to Persinger's interest in paranormal perception by such papers as "Putative Perception of Rotating Permanent Magnetic Fields Following Ingestion of LSD" (*Perceptual and Motor Skills,* vol. 87, 1998, pp. 601–602), in which Persinger claimed that students under the influence of LSD could "see" magnetic fields in a totally dark room.

103 the term *mystic* often refers: James, *The Varieties of Religious Experience,* p. 299.

PAGE

103 "supranormal" feats: *Cosmic Consciousness*, by Richard Bucke, Causeway Books, New York, 1974, p. 309.

notably Aldous Huxley and Huston Smith: Huxley took a credulous stance toward supernatural phenomena in *The Perennial Philosophy*, Harper & Row, New York, 1944, pp. 26–28; as did Smith in *Why Religion Matters*, HarperCollins, New York, 2001, pp. 219–220. Another book that virtually equated mysticism and parapsychology is *Psychiatry and Mysticism*, edited by Stanley Dean, Nelson-Hall, Chicago, 1975.

"I'm sure [psychic phenomena] exist": *Grace and Grit*, by Ken Wilber, Shambhala Publications, Boston, 1993, p. 390.

the cosmos is not "dead matter": Bucke, *Cosmic Consciousness*, p. 14.

104 the "causal operator": *Why God Won't Go Away*, by Andrew Newberg, Eugene D'Aquili, and Vince Rause, Ballantine Books, New York, 2001, p. 50.

people insist on finding a pattern: Ibid., p. 188.

"[I]n the face of chronic uncertainty": "Anthropomorphism: A Definition and a Theory," by Stewart Guthrie, in *Anthropomorphism, Anecdotes, and Animals*, edited by Robert Mitchell, Nicholas Thompson, and H. Lyn Miles, State University of New York Press, Albany, 1997, pp. 55–56. See also Guthrie's book *Faces in the Clouds*, Oxford University Press, New York, 1993; and his article "The Sacred: A Skeptical View," in *The Sacred and Its Scholars*, edited by Thomas Idinopulos and Edward Yonan, E. J. Brill, New York, 1996, pp. 124–138. Aldous Huxley lent support to the notion that mysticism stems from our anthropomorphism when he said in *The Perennial Philosophy* (p. 21) that "the Absolute Ground of all existence has a personal aspect."

105 "theory of mind" capacity: For a lucid review of the theory of mind theory, see *How the Mind Works*, by Steven Pinker, W. W. Norton, New York, 1997.

6. The Sheep Who Became a Goat

106 I met the British psychologist: I interviewed Susan Blackmore in Cambridge on April 29, 1999.

"fantasy-prone personality": I found this syndrome discussed in *The Three-Pound Universe*, by Judith Hooper and Dick Teresi, Jeremy Tarcher, Los Angeles, 1986, p. 344.

107 a spectacular out-of-body experience: Blackmore provided a detailed account of this experience and her subsequent attempts to understand out-of-body experiences in *Beyond the Body*, Academy Chicago Publishers, 1992. In this book, Blackmore pooh-poohed the myth that if you die in a dream, you die in reality. She once dreamed that she fell off a cliff and broke into "chunks" that were carried away by several ambulances. "I didn't seem to mind in the least" (pp. 109–110).

"I see it as perfectly logical": *In Search of the Light*, by Susan Blackmore, Prometheus Books, Amherst, New York, 1996, p. 270. In this charming autobio-

PAGE

graphical book, Blackmore told the story of her odyssey from psi believer to skeptic.

"hypnotic regression": See ibid., pp. 214–217.

ketamine: Blackmore described her reaction to ketamine in her book on near-death experiences, *Dying to Live,* Grafton, London, 1993, p. 176.

109 the theosophical concept of "thought forms": See Blackmore's discussion of theosophy and thought forms in *Beyond the Body.*

sheep . . . goats: The terms *sheep* and *goat* were coined in the 1940s by the American parapsychologist Gertrude Schmeidler, according to *Harper's Encyclopedia of Mystical and Paranormal Experience,* by Rosemary Ellen Guiley, Castle Books, Edison, N.J., 1991, p. 546.

110 "a number of errors in . . . the protocol": Blackmore, *In Search of the Light,* p. 130. See pp. 246–254 for a detailed account of this episode.

investigations of astral projection: Blackmore discussed astral projection and near-death experiences in "Near-Death Experiences: In or Out of the Body?," *Skeptical Inquirer,* Fall 1991, pp. 34–45.

a purely physiological explanation for tunnel visions: In ibid., Blackmore argued that her theory of tunnel vision is simpler than a similar theory proposed by the mathematical biologist Jack Cowan of the University of Chicago, which I mention in chapter 8.

111 why so many people . . . believe in telepathy: See Blackmore's articles "The Lure of the Paranormal," *New Scientist,* September 22, 1990, pp. 62–65; and "Psychic Experiences: Psychic Illusions," *Skeptical Inquirer,* Summer 1992, pp. 367–376.

But a statistical analysis shows: Blackmore mentioned this analysis, which was carried out by the statistician Christopher Scott, in her article "The Lure of the Paranormal," p. 63.

112 "Everywhere I have looked": Blackmore, *In Search of the Light,* p. 259.

"the science that will force": Ibid., p. 260. One of the most serious recent arguments for a new scientific paradigm that embraces psychic phenomena is *The Conscious Universe,* by Dean Radin, HarperSanFrancisco, 1997. For a critical review by a statistician of Radin's statistical arguments in favor of psi, see "Where Has the Billion Trillion Gone?," by I. J. Good, *Nature,* vol. 389, 1997, pp. 306–307.

113 "This is daft": Blackmore, *Beyond the Body,* p. 115.

"[J]ust as you become 'lucid' in a dream": Ibid., pp. 280–281. Also see Blackmore's articles "Dreams That Do What They're Told," *New Scientist,* January 6, 1990, pp. 48–51; and "Lucid Dreaming: Awake in Your Sleep?," *Skeptical Inquirer,* Summer 1991, pp. 362–370.

116 "waking from the meme dream": See "Waking from the Meme Dream," by Susan Blackmore, in *The Psychology of Awakening,* edited by Gay Watson, Stephen Batchelor, and Guy Claxton. "Who wakes up when the meme dream is all dismantled?" Blackmore asked. "Ah, there's a question" (p. 122).

PAGE

116 turned her skeptical eye on . . . meditation: Blackmore raised doubts about many of the claims made for meditation in "Is Meditation Good for You?," *New Scientist*, July 6, 1991, pp. 30–33.

117 "Freud-style skeptic about all things mystical": See "The Dazzling Dark," by John Wren-Lewis, in *What Is Enlightenment?*, Summer 1995, p. 40. In this article, Wren-Lewis told the story of his poisoning and subsequent awakening into permanent mystical awareness.

118 "will ever be entirely free of religion": *The Meme Machine*, by Susan Blackmore, Oxford University Press, New York, 1999, p. 201.
"Science is not perfect": Ibid., p. 202.

119 reality . . . "is just the evolutionary process": Ibid., p. 242.
"a recipe for selfishness and wickedness": Ibid., p. 245.

120 "radiant dark pure consciousness": Wren-Lewis, "The Dazzling Dark," p. 41.
"half wondering if the doctors had sawn part of it away": Ibid., p. 42.
pain became "simply an interesting sensation": Ibid.

121 "the very impulse to seek the joy": Ibid., p. 43.
Brain scans have tentatively linked: See "Depersonalization: Neurobiological Perspectives," by Mauricio Sierra and German Berrios, *Biological Psychiatry*, vol. 44, 1998, pp. 898–908.

122 "cause marked distress or impairment": *Diagnostic and Statistical Manual of Mental Disorders*, fourth edition, American Psychiatric Association, Washington, D.C., 1994, p. 488.

123 a Tibetan Buddhist practice called dream yoga: Dream yoga was described by the lucid-dreaming expert Stephen LaBerge in his book *Lucid Dreaming*, Ballantine Books, New York, 1985.

7. Zen and James Austin's Brain

124 "How really real": *Zen and the Brain*, by James Austin, MIT Press, Cambridge, 1998, p. 49.
*neo*realization and *neo*personalization: Ibid., p. 50.

125 "perennial psychophysiology": Ibid., p. xix.
Schizophrenics . . . mystical bliss: See ibid., pp. 30–34. See also "Hallucinogenic Drug-Induced States Resemble Acute Endogenous Psychoses: Results of an Empirical Study," by E. Gouzoulis-Mayfrank et al., *European Psychiatry*, vol. 13, 1998, pp. 399–406. This paper presented evidence that schizophrenics, especially in the early stages of their disease, occasionally experience "subjectively pleasant" states of "oceanic boundlessness."
"promethean hyperpraxia": Austin, *Zen and the Brain*, p. 674.

126 primarily right-brain phenomena: For a recent example of the linkage of the brain's right hemisphere to spirituality and mysticism, see *The Right Mind*, by Robert Ornstein, Harcourt Brace, New York, 1997.
chemical etching: Austin elaborated on this analogy in *Zen and the Brain*, pp. 653–659.

PAGE

127 practitioners of Transcendental Meditation were found to be sleeping: Austin, *Zen and the Brain,* p. 81. The paper cited by Austin was "Sleep During Transcendental Meditation," by R. Pagano et al., *Science,* vol. 191, 1976, pp. 308–309.
"Ultimate Pure Being": Austin, *Zen and the Brain,* pp. 627–630.
To reach him: I interviewed James Austin at his home on May 21, 2000.

128 During one meditation session: See *Zen and the Brain,* pp. 470–472, for Austin's account of this absorption experience.

129 "profound, implicit, perfect": Ibid., pp. 537–538.
"rinsed in a cloudburst . . . washed away": Ibid., p. 589.

132 He wore . . . a "stupid" smile: *An Anthropologist on Mars,* by Oliver Sacks, Vintage Books, New York, 1995, p. 44.

133 Greg's fate "could have been prevented": Ibid., p. 45.
"unseasoned 'new-age' mysticism": Austin, *Zen and the Brain,* p. 680.
"but it redeems its weaknesses": Ibid., p. 677.
[Bodhidharma's] legs became gangrenous: *Holy Madness,* by Georg Feuerstein, Paragon House, New York, 1990, p. 48.
tore his eyelids off: I heard this story about Bodhidharma and open-eyed meditation from a Zen teacher named Chisho Maas, from whom I have taken Zen lessons in my hometown.
chopped off his own hand: The Zen disciple who cut off his hand to demonstrate his seriousness to Bodhidharma was Hui-K'e, according to *The Three Pillars of Zen,* by Philip Kapleau, Anchor Books, Garden City, N.Y., 1980, p. 357.

134 a teacher slammed his door: Feuerstein, *Holy Madness,* p. 50.
the limbic system plays an *enhanced* role: See "The Neural Substrates of Religious Experience," by Jeffrey Saver and John Rabin, *Journal of Neuropsychiatry,* vol. 9, no. 3, 1997, pp. 498–510.
just as the Copernican revolution shattered: Austin made this analogy in *Zen and the Brain,* p. 47.

136 "most aspirants, perhaps authors in particular": Austin, *Zen and the Brain,* p. 295.
"far too long": "The Art of Letting Go," by Susan Blackmore, *Times* (London) *Higher Education Supplement,* May 7, 1999, p. 30.
"Knowledge is a process": Austin, *Zen and the Brain,* p. 641. Austin attributed the quote to Martin Fischer, a physician.
"Vast numbers of leads": Ibid., p. 696.
"study and predict some events": Ibid., p. 290.
"Rosetta stone": Ibid., p. 696.

137 the explanatory gap: This phrase was introduced in "Materialism and Qualia: The Explanatory Gap," by Joseph Levine, *Pacific Philosophical Quarterly,* vol. 64, 1983, pp. 354–361.
mysterianism: The term *mysterianism* is derived from the phrase "new mysterians," which the philosopher Owen Flanagan coined in *The Science of the Mind,* MIT Press, Cambridge, 1991, to refer to scientists and philosophers who doubt that consciousness can be scientifically explained.

PAGE

137 the "real challenge" for neuroscience: Austin, *Zen and the Brain*, p. 578.

138 instrumentality . . . automatization . . . deautomatization: Arthur Deikman has discussed these concepts in various publications, including "Deautomatization and the Mystic Experience," in *Altered States of Consciousness*, edited by Charles Tart, HarperCollins, New York, 1990, pp. 34–57; and "A Functional Approach to Mysticism," *Journal of Consciousness Studies*, vol. 7. no. 11–12, November/December 2000, pp. 75–92.

"Positive spiritual experiences": Austin, *Zen and the Brain*, p. 375.

139 doors of deception: This phrase was coined by my good friend and fellow ink-stained wretch Robert Hutchinson.

140 the age of scientifically based psychedelic mysticism: For an incisive history of psychedelic research over the past century, see "Psychiatric Research with Hallucinogens: What Have We Learned?," by Charles Grob, *The Heffter Review of Psychedelic Research*, vol. 1, 1998, pp. 8–20. Grob is a psychiatrist at the Harbor-UCLA Medical Center in Los Angeles.

8. In the Birthplace of LSD

141 "remarkable restlessness": *LSD: My Problem Child*, by Albert Hofmann, Jeremy Tarcher, Los Angeles, 1983, p. 15.

"furniture assumed . . . colored mask": Ibid., p. 17.

"I had not experimented thoughtlessly": Ibid., p. 18.

"the horror softened . . . newly created": Ibid., p. 19.

142 Bicycle Day: I learned about Bicycle Day from the psychologist Thomas Roberts.

effects of dimethyltryptamine: Stephen Szara first described DMT's effects in "Dimethyltryptamine: Its Metabolism in Man: The Relation of Its Psychotic Effect to the Serotonin Metabolism," *Experientia*, vol. 12, no. 11, 1956, pp. 441–444.

investigations of psychedelics helped reveal: See *Drugs and the Brain*, by Solomon Snyder, Scientific American Books, New York, 1986, p. 205, in which the author, a leading neuroscientist, describes how research on LSD and other psychedelics helped neuroscientists understand how serotonin and other neurotransmitters work in the brain. For an irreverent but well-reported account of the history of psychedelic research, see *Trips*, by Cheryl Pellerin, Seven Stories Press, New York, 1998. One of the best books on the social aspects of the psychedelic era is *Storming Heaven*, by Jay Stevens, Grove Press, New York, 1987.

one thousand papers . . . forty thousand patients: *Psychedelic Drugs Reconsidered*, by Lester Grinspoon and James Bakalar, The Lindesmith Center, New York, 1997, p. 192.

a book by two journalists: *LSD: The Problem-Solving Psychedelic*, by P. G. Stafford and B. H. Golightly, Award Books, New York, 1967.

143 there has been a quiet resurgence: For a recent report on research into the

PAGE

therapeutic benefits of psychedelics, see "Scientists Test Hallucinogens for Mental Ills," by Sandra Blakeslee, *New York Times,* March 13, 2001, p. F1.

Heffter Research Institute: Information on the institute can be found at www.heffter.org. James Thornton, its executive director,was a high-ranking attorney for the Natural Resources Defense Council, a major environmental organization, when he underwent an identity crisis. In his moving, elegant book *A Field Guide to the Soul* (Bell Tower, New York, 1999), he told the story of how he became disillusioned with environmental activism, quit his job, studied under various spiritual teachers, and eventually found solace in a nature-based mysticism.

144 MAPS*:* Information on MAPS can be found at www.maps.org.

Doblin has also examined: "Dr. Leary's Concord Prison Experiment: A 34-Year Follow-Up Study," by Rick Doblin, *Journal of Psychoactive Drugs,* vol. 30, no. 4, 1998, pp. 419–426.

"We had kept twice as many convicts": Ibid., p. 420.

"Leary's misleading reports": Ibid., p. 425.

145 "Worlds of Consciousness": The Third International Congress of the European College for the Study of Consciousness (whose translated acronym is ECBS) was held in Basel, Switzerland, November 11–14, 1999. The ECBS Web site is www.magnet.ch/ecbs.

Salvia divinorum: This hallucinogenic plant has recently become faddish. See "New Cautions over a Plant with a Buzz," by Richard Lezon Jones, *New York Times,* July 9, 2001, p. B1. But the article quoted one supplier of the herb saying that only one in ten customers of *Salvia divinorum* comes back for more.

"gentle yet penetrating touches": *Interior Castle,* by Saint Teresa of Ávila, Image Books, Garden City, N.Y., 1961, p. 222.

146 as Nichols and . . . Shulgin first pointed out: "Characterization of Three New Psychotomimetics," by David Nichols and Alexander Shulgin, in *The Pharmacology of Hallucinogens,* edited by R. C. Stillman and R. E. Wilette, Pergamon Press, New York, 1978, p. 74.

Nichols coined the term *entactogen:* "Differences Between the Mechanism of Action of MDMA, MBDB, and the Classical Hallucinogens: Identification of a New Therapeutic Class: Entactogens," by David Nichols, *Journal of Psychoactive Drugs,* vol. 18, 1986, pp. 305–318. See also Nichols's recent review of psychedelic research, "From Eleusis to PET Scans: The Mysteries of Psychedelics," *MAPS,* Winter 1999–2000, pp. 49–55.

"few, if any, long-term neuropsychological deficits": "Do Hallucinogens Cause Residual Neurophysiological Toxicity?," by John Halpern and Harrison Pope, *Drug and Alcohol Dependence,* vol. 53, 1999, p. 247.

147 psychedelics . . . should be reconsidered: "The Use of Hallucinogens in the Treatment of Addiction," by John Halpern, *Addiction Research,* vol. 4, no. 2, 1996, pp. 177–189.

treated alcoholics with ketamine: See "Ketamine Psychedelic Therapy (KPT): A

PAGE

Review of the Results of Ten Years of Research," by Evgeny Krupitsky and A. Y. Grinenko, *Journal of Psychoactive Drugs,* vol. 29, no. 2, 1997, pp. 165–183. Krupitsky has also administered ketamine therapy to heroin addicts. See "Ketamine-Assisted Psychotherapy of Heroin Addiction," by Evgeny Krupitsky et al., *MAPS,* Winter 1999–2000, pp. 21–26. Ketamine produces effects remarkably similar to near-death experiences, according to the British psychiatrist Karl Jansen. In *Ketamine: Dreams and Realities,* MAPS, Sarasota, Florida, 2001, Jansen argued that endogenous (naturally occurring) neurochemicals similar to ketamine might play a role in near-death experiences. His theory is similar to the endogenous-DMT theory of the psychiatrist Rick Strassman, described in chapter 9.

148 "All I want is out of here": I found this account of Burroughs's ayahuasca adventure in *One River,* by Wade Davis, Touchstone, New York, 1996, p. 155. Burroughs's correspondence with Allen Ginsberg was published as *The Yage Letters,* City Lights Books, San Francisco, 1963.

botanist Richard Shultes: Shultes died on April 10, 2001, at the age of eighty-six. For an authoritative, affectionate, gripping account of his career by one of his former students, see Davis, *One River.*

149 examination of members of the União do Vegetal: This study's results were presented in "Human Psychopharmacology of Hoasca: A Plant Hallucinogen Used in Ritual Context in Brazil," by Charles Grob et al., *Journal of Nervous and Mental Disease,* vol. 184, no. 2, pp. 86–94. The study compared fifteen members of the União do Vegetal to fifteen controls.

149 UDV members had elevated levels of serotonin: "Platelet Serotonin Uptake Sites Increased in Drinkers of *Ayahuasca,*" by James Callaway et al., *Psychopharmacology,* vol. 116, 1994, pp. 385–387.

[Ratsch's] books and articles: Ratsch edited and contributed two articles to an English-language book on entheogenic plants: *Gateway to Inner Space,* Avery Publishing, Garden City Park, N.Y., 1989.

151 Vollenweider has tentatively confirmed: "Recent Advances and Concepts in the Search for Biological Correlates of Hallucinogen-Induced Altered States of Consciousness," by Franz Vollenweider, *The Heffter Review of Psychedelic Research,* vol. 1, 1998, pp. 21–32. (See the citations for Adolf Dittrich in the notes to the introduction.) In addition to Dittrich, another researcher who has influenced Vollenweider is the Swiss psychiatrist Christian Scharfetter, who proposed a theory of the ego that I find obscure but that helped Vollenweider understand pathological and benign altered states. See "Ego-Psychopathology: The Concept and Its Evaluation," by Christian Scharfetter, *Psychological Medicine,* vol. 11, 1981, pp. 273–280. Scharfetter said that schizophrenics may become agitated if they meditate by themselves, but they may be soothed if they sit with someone else who is meditating.

152 psilocybin helps subjects tap into: See "Increased Activation of Indirect Seman-

PAGE

tic Associations under Psilocybin," by Manfred Spitzer et al., *Biological Psychiatry,* vol. 39, 1996, pp. 1055–1057.

153 long-term use of MDMA: Vollenweider presented his views on MDMA toxicity in "Is a Single Dose of MDMA Harmless?," *Neuropsychopharmacology,* vol. 21, no. 4, 1999, pp. 598–600; and "Psychological and Cardiovascular Effects and Short-Term Sequelae of MDMA ('Ecstasy') in MDMA-Naive Healthy Volunteers," by Franz Vollenweider et al., *Neuropsychopharmacology,* vol. 19, no. 4, 1998, pp. 241–251.

pre-pulse inhibition: See "Opposite Effects of 3,4-Methylenedioxymethamphetamine (MDMA) on Sensorimotor Gating in Rats Versus Healthy Humans," by Franz Vollenweider et al., *Psychopharmacology,* vol. 143, 1999, pp. 365–372.

154 psychedelics are psychotomimetics: See "Psilocybin Induces Schizophrenia-like Psychosis in Humans via a Serotonin-2 Agonist Action," by Franz Vollenweider et al., *Cognitive Neuroscience,* vol. 9, no. 17, 1998, pp. 3897–3902; and "Positron Emission Tomography and Fluorodeoxyglucose Studies of Metabolic Hyperfrontality and Psychopathology in the Psilocybin Model of Psychosis," by Franz Vollenweider et al., *Neuropsychopharmacology,* vol. 16, no. 5, 1997, pp. 357–372. Vollenweider's suggestion that psychedelic states resemble early-onset psychosis is corroborated by "Hallucinogenic Drug–Induced States Resemble Acute Endogenous Psychoses: Results of an Empirical Study," by E. Gouzoulis-Mayfrank et al., *European Psychiatry,* vol. 13, 1998, pp. 399–406.

156 [Hofmann's] worst trip: His horrific psilocybin trip took place in 1962 at the home of the German writer Ernst Junger. Hofmann recalled the trip in *LSD: My Problem Child,* pp. 162–168.

"Only when we are conversant": "LSD: Completely Personal," by Albert Hofmann, *MAPS,* Summer 1996, pp. 51–52.

"represent a forbidden transgression": Hofmann, *LSD: My Problem Child,* p. 157.

"God is a substance": Ibid., p. 159.

157 synesthesia . . . has been estimated: I found this statistic in a book review, ". . . But What Does 'Blue' Smell Like?," by Ilya Farber, *Nature,* April 12, 2001, p. 744. Farber, in turn, cited the book he was reviewing, *Synesthesia: The Strangest Thing,* by John Harrison, Oxford University Press, Oxford, 2001. For an interesting recent attempt to unravel the cognitive pathways leading to synesthesia, see "Unconscious Priming Eliminates Automatic Binding of Colour and Alphanumeric Form in Synesthesia," by Jason Mattingly et al., *Nature,* March 29, 2001, pp. 580–582.

synesthesia . . . might result from short-circuiting: I found this hypothesis in Snyder, *Drugs and the Brain,* p. 203.

Cowan's theory depicts the brain: "Geometric Visual Hallucinations, Euclidean Symmetry, and the Functional Architecture of Striate Cortex," by Paul Bresslof et al., *Philosophical Transactions of the Royal Society of London,* Series B, vol. 356,

PAGE

2001, pp. 299–330. For a more recent journalistic account of Cowan's work, see "Secrets of an Acidhead," by Dana Mackenzie, *New Scientist,* June 23, 2001, pp. 26–30. Susan Blackmore's explanation of tunnel hallucinations, described in chapter six, is a modified version of Cowan's.

158 even pigeons perceive form constants: *Intoxication,* by Ronald Siegel, E. P. Dutton, New York, 1989, pp. 165–166. In this book, Siegel describes many other drug-related experiments he performed on humans and animals.

"many in the nature of cartoons": "Hallucinations," by Ronald Siegel, *Scientific American,* October 1977, p. 133.

9. God's Psychoanalyst

160 transpersonal psychology: Grof helped to organize early conferences on transpersonal psychology and to found two of its primary institutions: the Association of Transpersonal Psychology, based in Palo Alto, California, which publishes the *Journal of Transpersonal Psychology;* and the International Transpersonal Association, which has hosted more than fifteen conferences around the world.

"forefather of transpersonal psychology": *Psychology of the Future,* by Stanislav Grof, State University of New York Press, Albany, 2000, p. 289.

161 Grof and his second wife: Grof's first wife was the anthropologist turned spiritual leader Joan Halifax, who for years was Grof's partner in administering LSD psychotherapy.

holotropic breathwork: For a first-person account by a journalist who underwent holotropic breathwork training, see *What Really Matters,* by Tony Schwartz, Bantam Books, New York, 1995, pp. 42–51. Schwartz also presented profiles of Stanislav Grof, Ken Wilber, and other mystical intellectuals.

"dramatic improvement" in an obsessive-compulsive: *LSD Psychotherapy,* by Stanislav Grof, Hunter House, Alameda, Calif., 1980, p. 201. As Grof noted in this book, there were two major forms of LSD therapy. So-called psycholytic therapy resembled conventional psychotherapy except that the patient was under the influence of small doses of LSD, from 50 to 100 micrograms. Psychedelic therapy, which was Grof's specialty, called for doses of 300 micrograms or more, usually more than enough to propel the patient into transpersonal realms; the patient would often lie blindfolded on a couch or bed. Whereas psycholytic therapy typically comprised biweekly drug sessions lasting a year or more, psychedelic therapy generally involved only a few drug sessions supplemented by conventional psychotherapy sessions.

LSD followed by electroshock therapy: Ibid., p. 31.

Bluebird and MK-ULTRA: For accounts of these CIA programs, see *The Search for the "Manchurian Candidate,"* by John Marks, W. W. Norton, New York, 1991; and *Acid Dreams,* by Martin Lee and Bruce Shlain, Grove Press, New York, 1992.

LSD's "negative potential": Grof, *LSD Psychotherapy,* p. 13.

PAGE

162 greeting me at the door: I interviewed Grof on August 21, 1999.

167 He claimed that a colleague, Richard Tarnas: Richard Tarnas, who, like Grof, teaches at the California Institute of Integral Studies, is the author of a history of Western thought titled *The Passion of the Western Mind,* Ballantine Books, New York, 1991. Tarnas has been working for years on a book that attempts to rehabilitate astrology. Grof has cited Tarnas's astrology research in *LSD Psychotherapy,* p. 243; and, more recently, in *Psychology of the Future,* p. 242.

Blackmore found that people delivered by cesarean section: See "Near-Death Experiences: In or Out of the Body?," by Susan Blackmore, *Skeptical Inquirer,* Fall 1991, pp. 38–39.

168 "is exposed . . . *Time, Life, Newsweek*": *The Varieties of Psychedelic Experience,* by Robert Masters and Jean Houston, Holt, Rinehart & Winston, New York, 1966, p. 306.

169 addressing ancient theological riddles: Two of Grof's explicitly theological works are *The Cosmic Game,* State University of New York Press, Albany, 1998; and *Psychology of the Future.*

171 "principle of maximum diversity": *Infinite in All Directions,* by Freeman Dyson, Harper & Row, New York, 1988, p. 298. I discussed Dyson's concept with him in *The End of Science,* pp. 251–255.

172 Buddhaverse: The Tibetan Buddhism scholar Robert Thurman introduced this term in his book *Inner Revolution,* Riverhead Books, New York, 1998.

[Christina] feared she was losing her mind: I found this account of Christina Grof's travails in "Two Lives, One Path," by Keith Thompson, *Noetic Sciences Review,* twenty-fifth anniversary issue, 1998, pp. 39–56.

"spiritual emergencies": Stanislav and Christina Grof edited a book of essays on this topic: *Spiritual Emergency,* Jeremy Tarcher, Los Angeles, 1989.

173 Grof found it "implausible": Grof, *Psychology of the Future,* p. 167.

"manifestations of archetypal elements": Ibid., p. 168.

174 speculation that endogenous DMT contributes to schizophrenia: Findings on endogenous DMT were summarized in "The Psychedelic Model of Schizophrenia: The Case of N,N-Dimethyltryptamine," by J. Christian Gillin et al., *American Journal of Psychiatry,* vol. 133, 1976, pp. 203–208; and "N,N-Dimethyltryptamine: An Endogenous Hallucinogen," by Steven Barker et al., *International Review of Neurobiology,* vol. 22, 1981, pp. 83–110. These and other articles on DMT were cited by Rick Strassman in *DMT: The Spirit Molecule,* Park Street Press, Rochester, Vt., 2001. Among the many peer-reviewed articles that Strassman has written on psychedelics are "Adverse Reactions to Psychedelic Drugs," *Journal of Nervous and Mental Disease,* vol. 172, no. 10, 1984, pp. 577–595; "Hallucinogenic Drugs in Psychiatric Research and Treatment: Perspectives and Prospects," *Journal of Nervous and Mental Disease,* vol. 183, no. 3, 1995, pp. 127–138; and "Endogenous Ketamine-like Compounds and the NDE: If So, So What?," *Journal of Near-Death Studies,* Fall 1997, pp. 27–41. In this final article, Strassman cri-

PAGE

tiqued the proposal of Karl Jansen that endogenous neurochemicals similar to the anesthetic ketamine might trigger near-death experiences and argued that endogenous DMT secretions are a more likely cause.

175 "Who ordered that?": I found this quote from I. I. Rabi in *Lonely Hearts of the Cosmos,* by Dennis Overbye, HarperPerennial, New York, 1992, p. 201.

many cultures have purveyed legends: *The Meme Machine,* by Susan Blackmore, Oxford University Press, New York, 1999, pp. 176–178.

176 "a powerful sense that there is somebody": Ibid., p. 176.

"Given how common this syndrome is": *Phantoms in the Brain,* by V. S. Ramachandran and Sandra Blakeslee, William Morrow, New York, 1998, p. 106. See also pp. 85–87, where Ramachandran speculated that the cartoonist and author James Thurber, who suffered from an eye disease that blinded him in middle age, may have had Charles Bonnet syndrome.

10. The Man in the Purple Sparkly Suit

177 one of his free-form talks: I saw Terence McKenna perform in New York City on May 6, 1999, under the auspices of the Open Center.

"the Timothy Leary of the 90's": "Terence McKenna, 53, Dies: Patron of Psychedelic Drugs," by Douglas Martin, *New York Times,* April 9, 2000, p. 40.

"a clumsy word freighted": *Food of the Gods,* by Terence McKenna, Bantam Books, New York, 1992, p. 112.

As Strassman has acknowledged: "More than anyone," Strassman wrote in *DMT: The Spirit Molecule,* "McKenna has raised awareness of DMT, through lectures, books, interviews, and recordings, to its present unprecedented level" (p. 349).

"self-transforming machine elves": McKenna, *Food of the Gods,* p. 258.

"interdimensional nexus": Ibid., p. 261.

178 a book of wild psychedelic speculation: *The Invisible Landscape,* by Terence McKenna and Dennis McKenna, Seabury Press, New York, 1975.

earliest known art . . . form constants: Recent research on form constants, primitive art, and psychedelics was discussed in "Ancient Altered States," by Mary Roach, *Discover,* June 1998, pp. 52–58.

179 "the tradition of healing, divination": McKenna, *Food of the Gods,* p. 4.

"For the shaman, the cosmos is a tale": Ibid., p. 7.

181 "done nothing to mitigate": *True Hallucinations,* by Terence McKenna, HarperSanFrancisco, 1993, p. 226.

183 superstrings . . . dead end: For a recent critique by a physicist, see "Is String Theory Even Wrong?," by Peter Woit, *American Scientist,* March/April 2000, pp. 110–112.

184 "the only phosphoric acid–containing indole in all of nature": Albert Hofmann drew attention to this peculiar feature of psilocybin in *LSD: My Problem Child,* Jeremy Tarcher, Los Angeles, 1983, p. 114.

the Harvard psychiatrist John Mack: Mack made the case for alien abductions

PAGE

in *Abductions: Human Encounters with Aliens,* Charles Scribner's Sons, New York, 1994.

188 "neophilia": *Forgotten Truth,* by Huston Smith, HarperSanFrancisco, 1992, p. 145.

189 our reality might actually be virtual reality: I discussed the belief of various scientists that reality might be — or become — a computer simulation in *The End of Science,* Broadway Books, New York, 1997, pp. 247–260.

190 "I shall pass from thousands of apparitions": "The Zahir," by Jorge Luis Borges, *A Personal Anthology,* Grove Press, New York, 1967, p. 137.

191 "There are various options . . . think about it all": "Terence McKenna's Last Trip," by Erik Davis, *Wired,* May 2000, p. 162.

"Some people would certainly argue": Ibid., pp. 166–168.

192 to McKenna wonder was the essence: McKenna's fondness for wonder may have explained his fondness for DMT. In *Food of the Gods,* he said that under the influence of DMT one has "a sense of other times, and of one's own infancy, and of wonder, wonder, and more wonder" (p. 258). In *True Hallucinations,* he said: "I am occasionally asked if DMT is dangerous. The proper answer is that it is only dangerous if you feel threatened by the possibility of death by astonishment" (p. 8).

"In the gnostic conception . . . abyss": *The Gnostic Religion,* by Hans Jonas, revised second edition, Beacon Press, Boston, 1991, pp. 338–339.

193 "remarkably happy . . . free will for tumors?": "A Designer Universe?," in *Facing Up,* by Steven Weinberg, Harvard University Press, Cambridge, 2001, p. 240. This essay was originally published in the *New York Review of Books,* October 21, 1999, pp. 46–48. I profiled Weinberg in *The End of Science,* pp. 71–77.

"The notion of some kind of": McKenna, *True Hallucinations,* p. 202.

194 "final time" . . . "end of the wave": Ibid., pp. 198, 200.

The obituary was for Jeffrey Willick: "Jeffrey Willick, 40, Astrophysicist and Professor," by Robert Hanley, *New York Times,* June 20, 2000, p. B9.

11. Ayahuasca

196 Shulgin is credited with: Shulgin and his coauthor, David Nichols, professor of medicinal chemistry at Purdue University, described MDMA's effects in "Characterization of Three New Psychotomimetics," in *The Pharmacology of Hallucinogens,* edited by R. C. Stillman and R. E. Wilette, Pergamon Press, New York, 1978, p. 74.

197 "Hand in the air": *PIHKAL,* by Alexander Shulgin and Ann Shulgin, Transform Press, Berkeley, Calif., 1991, p. xxvii.

"Of course, if an established couple": Ibid., p. xxviii.

"No one who is lacking": Ibid., p. xi.

198 "the agony and ecstasy of ayahuasca": *TIHKAL,* by Alexander Shulgin and Ann Shulgin, Transform Press, Berkeley, Calif., 1997, p. 304.

"incredibly complex and insane roads": Ibid., p. 305.

PAGE

198 "Oh no," he thought: Ibid., p. 306.
We sat in a living room: I interviewed the Shulgins on August 19, 1999.

201 Sasha said his view of death: Sasha discussed how he came to terms with his old age and mortality in *PIHKAL*, pp. 419–427.

205 It tasted astonishingly foul: In his book *One River* (Touchstone, New York, 1996, p. 191), the anthropologist Wade Davis said that ayahuasca smells and tastes like "the entire jungle ground up and mixed with bile."

208 "merry elfin . . . eggs on the rebound": *True Hallucinations*, by Terence McKenna, HarperSanFrancisco, 1993, p. 7.
through some strange synesthesia: Ibid., p. 67.

209 Classic DMT hallucinations, Kevin said: I found many features of my ayahuasca trip reflected in the descriptions of DMT visions in *DMT: The Spirit Molecule*, by Rick Strassman, Park Street Press, Rochester, Vt., 2001.

12. The Awe-ful Truth

214 Greeting me at the door of his home: I interviewed Huston Smith in Berkeley on March 25, 2000.

215 Smith's formulation of mysticism: Smith discussed mystical awe, Rudolf Otto, and other related issues in *Cleansing the Doors of Perception*, Jeremy Tarcher/Putnam, New York, 2000. On p. 12, he quoted this passage from *Duino Elegies* by Rainer Maria Rilke (translated by Stephen Mitchell): "Beauty is nothing / but the beginning of terror, which we still are just able to endure, / and we are so awed because it serenely disdains / to annihilate us."
"chill and numb": *The Idea of the Holy*, by Rudolf Otto, Oxford University Press, New York, 1958, p. 28.
"an almost grisly horror": Ibid., p. 13.

216 "wholly other . . . the opposite of everything": Ibid., pp. 29–30.
"They are attempts . . . frankly excluded": Ibid., pp. 26–27.

217 "a staggering mystery": *The Marriage of Sense and Soul*, by Ken Wilber, Broadway Books, New York, 1998, p. xi.

218 why creation occurred in the first place: For a review of cosmological theories of creation by an eminent astronomer, see *Before the Beginning*, by Martin Rees, Addison-Wesley, Reading, Mass., 1997.
"No one is certain what happened": Steven Weinberg made this statement in "Before the Big Bang," an essay in *Facing Up*, Harvard University Press, Cambridge, 2001, p. 165.

219 the Goldilocks dilemma: For an overview of the Goldilocks problem, see *In the Beginning*, by John Gribbon, Little, Brown, Boston, 1993, pp. 158–187.
the physicist Lawrence Krauss has calculated: I first cited this calculation by Krauss in my article "Universal Truths," *Scientific American*, October 1990, p. 112.
the anthropic principle: For a critique of the anthropic principle, see "Anthropic

PAGE

Design: Does the Cosmos Show Evidence of Purpose?," by Victor Stenger, *Skeptical Inquirer,* July/August 1999, pp. 40–43.

life "is a mystery no longer": *The Blind Watchmaker,* by Richard Dawkins, W. W. Norton, New York, 1986, p. ix.

life must be a ubiquitous phenomenon: For expositions of this thesis, see *Are We Alone?,* by Paul Davies, Basic Books, New York, 1995; and *Probability 1: The Book That Proves There Is Life in Outer Space,* by Amir Aczel, Harcourt Brace, New York, 2000. For a strong counterargument, see *Rare Earth,* by Peter Ward and Donald Brownlee, Springer-Verlag, New York, 2000, in which a paleontologist (Ward) and an astronomer (Brownlee) argue that multicellular organisms may be extremely rare in the universe and unique to earth. For a summary of Ward and Brownlee's arguments, see "Maybe We *Are* Alone in the Universe, After All," by William Broad, *New York Times,* February 8, 2000, p. F1.

220 fossilized microbes in a meteorite: See "Requiem for Life on Mars? Support for Microbes Fades," by Richard Kerr, *Science,* November 20, 1998, pp. 1398–1400.

"the origin of life appears to be": *Life Itself,* by Francis Crick, Simon & Schuster, New York, 1981, p. 88.

Stephen Jay Gould has estimated: In *Wonderful Life* (W. W. Norton, New York, 1989, p. 289), Gould declared: "Replay the tape a million times . . . and I doubt that anything like *Homo sapiens* would ever evolve again. It is, indeed, a wonderful life." On p. 14, Gould called evolution "a staggeringly improbable series of events, sensible enough in retrospect and subject to rigorous explanation but utterly unpredictable and quite unrepeatable." On the other hand, Gould asserted — inconsistently, in my view — that the origin of life on earth "was virtually inevitable, given the chemical composition of early oceans and atmospheres, and the physical principles of self-organizing systems" (p. 289).

SETI . . . is futile: See "The Search for Extraterrestrial Life," by Ernst Mayr, in *Extraterrestrials: Where Are They?,* second edition, edited by Ben Zuckerman and Michael Hart, Cambridge University Press, New York, 1995, pp. 152–156.

221 "Fundamental science may be": *The Creation,* by Peter Atkins, W. H. Freeman, San Francisco, 1981, p. 126.

Atkins envisioned science as a universal acid: See "Will Science Ever Fail?," by Peter Atkins, *New Scientist,* August 8, 1992, pp. 32–35.

"Complete knowledge . . . sunrise": Atkins, *The Creation,* p. 127.

"Our sense of wonder grows exponentially": *Biophilia,* by Edward O. Wilson, Harvard University Press, Cambridge, 1984, p. 10.

five compelling reasons: "Celebrating Creation," by Chet Raymo, *Skeptical Inquirer,* July/August 1999, pp. 21–23.

222 "We are contingent, ephemeral": *Skeptics and True Believers,* by Chet Raymo, Walker and Company, New York, 1998, p. 245.

"there is no way that a happy ending": *Why Religion Matters,* by Huston Smith, HarperSanFrancisco, 2001, p. 36.

PAGE

222 Dyson has calculated that consciousness: Dyson first presented his proposals on endless life in "Time Without End: Physics and Biology in an Open Universe," *Reviews of Modern Physics*, vol. 51, 1979, pp. 447–460. He offered a nonmathematical version of these ideas in "How Will It All End?," in *Infinite in All Directions*, Harper & Row, New York, 1988, pp. 97–121. Dyson's speculations inspired other cosmologists to ponder life's ultimate fate in an expanding universe. For an update on these speculations, see "The End of Everything," by Dennis Overbye, *New York Times*, January 1, 2002, p. F1.

"The more the universe seems": *The First Three Minutes*, by Steven Weinberg, Basic Books, New York, 1977, p. 154.

"Omega Point": The chief proponent of the Omega Point theory is the physicist Frank Tipler. See his book *The Physics of Immortality*, Doubleday, New York, 1994.

"scientific theology": I coined this term and discussed its significance in *The End of Science*, Broadway Books, New York, 1997, pp. 247–260. In these pages I also presented profiles of Freeman Dyson, Frank Tipler, and other practitioners of scientific theology.

224 ladder that we should "throw away": *Tractatus Logico-Philosophicus*, by Ludwig Wittgenstein, Routledge, New York, 1990, p. 189.

225 "It is truly extravagant to define God": I found Voltaire's quote in *The Story of Philosophy*, by Will Durant, Simon & Schuster, New York, 1961, p. 177.

"enriching the wonder of normality": *Psychedelic Drugs Reconsidered*, by Lester Grinspoon and James Bakalar, The Lindesmith Center, New York, 1997, p. 288.

"beyond its present limitations": *The Varieties of Psychedelic Experience*, by Robert Masters and Jean Houston, Holt, Rinehart & Winston, New York, 1966, p. 316.

"I am by nature not pro-drug": Jean Houston made this remark when I interviewed her at her home on September 2, 1999.

226 "People would get addicted to it": *A Mythic Life*, by Jean Houston, HarperSanFrancisco, 1996, p. 196.

"Attention. Attention": This story is told in *The Three Pillars of Zen*, by Philip Kapleau, Anchor Books, Garden City, N.Y., 1980, p. 10. The scholar Gerald Heard once wrote: "It is the unique quality of attention which LSD can bestow that will or will not be of benefit" (*The Psychedelic Reader*, edited by Timothy Leary, Ralph Metzner, and Gunther Weil, Carol Publishing, New York, 1993, p. 10).

227 "snap out of the movie of life": *One Taste*, by Ken Wilber, Shambhala Publications, Boston, 1999, p. 60. Similarly, Roger Walsh, a transpersonal psychiatrist at the University of California at Irvine and a close friend of Wilber's, likened enlightenment to lucid dreaming when I interviewed him and his wife, Frances Vaughan, a psychologist, in Mill Valley, California, on August 20, 1999.

228 "the overcoming . . . stop and think": *The Varieties of Religious Experience*, by William James, Macmillan, New York, 1961, p. 329.

PAGE

"alter one's relationship": *The Guru Papers,* by Joel Kramer and Diana Alstad, Frog, Ltd., Berkeley, Calif., 1993, p. 316.

229 "has within it a hidden duality": Ibid., p. 306.

"The very nature of": Ibid., p. 311.

enlightenment . . . unattainable by mere mortals: Carl Jung presented an eloquent argument against the possibility of total enlightenment in "Answer to Job." In the essay's final sentence, Jung asserted that "even the enlightened person remains what he is, and is never more than his own limited ego before the One who dwells within him, whose form has no knowable boundaries, who encompasses him on all sides, fathomless as the abysms of the earth and vast as the sky" (*The Portable Jung,* edited by Joseph Campbell, Penguin Books, New York, 1971, p. 650). Statements like this no doubt led Ken Wilber to argue that Jung was not a true mystic (*Grace and Grit,* Shambhala Publications, Boston, 1991, pp. 179–182).

230 "'wrong' actions won't arise": *Zen and the Brain,* by James Austin, MIT Press, Cambridge, 1998, p. 646.

"When all things return to the one": Ibid., p. 107.

"I want to taste sugar": As quoted in Smith, *Why Religion Matters,* p. 270.

"The Mirror of Madmen": I found Chesterton's poem in *Beyond Theology,* by Alan Watts, Meridian Books, New York, pp. 210–211. The first stanza reads: "I dreamed a dream of heaven, white as frost, / The splendid stillness of a living host; / Vast choirs of upturned faces, line o'er line. / Then my blood froze; for every face was mine." At the end of the poem, the narrator wakes up in a tavern and sees "The sight of all my life most full of grace, / A gin-damned drunkard's half-witted face." Watts, a hip, British-born philosopher, wrote that Chesterton here confused the egotistical, mortal, individual self with the eternal, infinite Self of which mystics speak.

231 "God waxes and wanes": "Creatures Seek God Through Me," by Meister Eckhart, *Meister Eckhart,* Harper & Row, 1941, p. 225. For a fascinating discussion of the problematic aspects of mystical identification with nature as a literary theme, see *Candor and Perversion,* by Roger Shattuck, W. W. Norton, New York, 1999, pp. 43–53. Shattuck, a distinguished literary critic, asked, "[W]here do we find significant criticism of or an opposition to the belief in the interconnectedness of all things?" (p. 49). In *Forbidden Knowledge,* Harcourt Brace, New York, 1996, Shattuck also wrote compellingly about the limits of human scientific, philosophical, literary, and religious knowledge.

"Omega race of humane beings": Austin, *Zen and the Brain,* p. 687.

Shulgin has defined a plus-four experience: *PIHKAL,* by Alexander Shulgin and Ann Shulgin, Transform Press, Berkeley, Calif., 1991, p. 964.

233 I have no choice but to choose: I have borrowed this phrasing from William James, who wrote, "My first act of free will shall be to believe in free will" (as quoted in "William James and the Case of the Epileptic Patient," by Louis Menand, *New York Review of Books,* December 17, 1998, p. 82).

PAGE

233 "convictionally impaired": Smith, *Why Religion Matters,* p. 91.

Epilogue: Winter Solstice

235 "My whole fucking life": *Grace and Grit,* by Ken Wilber, Shambhala Publications, Boston, 1991, p. 309.

"I would like to claim": Ibid., p. 311.

Buddha . . . leaving his wife and child: I am indebted to Diana Alstad, coauthor of *The Guru Papers,* for drawing my attention to this fact of Buddha's life.

Selected Bibliography

Andreson, Jensine, and Robert Forman, editors, *Cognitive Models and Spiritual Maps,* Imprint Academic, Bowling Green, Ohio, 2001.
Angel, Leonard, *Enlightenment East & West,* State University of New York Press, Albany, 1994.
Armstrong, Karen, *A History of God,* Ballantine Books, New York, 1993.
Austin, James, *Chase, Chance & Creativity,* Columbia University Press, New York, 1978.
———, *Zen and the Brain,* MIT Press, Cambridge, 1998.
Batchelor, Stephen, *The Faith to Doubt,* Parallax Press, Berkeley, California, 1990.
———, *Buddhism Without Beliefs,* Riverhead Books, New York, 1997.
Blackmore, Susan, *Beyond the Body,* Academy Chicago Publishers, 1992.
———, *Dying to Live,* HarperCollins, London, 1993.
———, *In Search of the Light,* Prometheus Books, Amherst, New York, 1996.
———, *The Meme Machine,* Oxford University Press, New York, 1999.
Borges, Jorges Luis, *A Personal Anthology,* Grove Press, New York, 1967.
Brothers, Leslie, *Mistaken Identity,* State University of New York Press, Albany, 2001.
Bucke, Richard, *Cosmic Consciousness,* Causeway Books, 1974.
Capra, Fritjof, *The Tao of Physics,* Shambhala Publications, Boston, 1975.
Davis, Wade, *One River,* Touchstone, New York, 1996.
DeKorne, Jim, *Psychedelic Shamanism,* Loompanics Unlimited, Port Townsend, Washington, 1994.
Dyson, Freeman, *Infinite in All Directions,* Harper & Row, New York, 1988.
Feuerstein, Georg, *Holy Madness,* Paragon House, New York, 1991.
Fiddes, Paul, *The Creative Suffering of God,* Oxford University Press, Oxford, 1992.
Forman, Robert, *Mysticism, Mind, Consciousness,* State University of New York Press, Albany, 1999.
———, editor, *The Problem of Pure Consciousness,* Oxford University Press, New York, 1990.

———, editor, *The Innate Capacity*, Oxford University Press, New York, 1998.

Forte, Robert, editor, *Entheogens and the Future of Religion*, Council on Spiritual Practices, San Francisco, 1997.

Gardner, Howard, *Intelligence Reframed*, Basic Books, New York, 1999.

Goodman, Martin, *I Was Carlos Castaneda*, Three Rivers, New York, 2001.

Greeley, Andrew, *Ecstasy: A Way of Knowing*, Englewood Cliffs, New Jersey, Prentice-Hall, 1974.

Green, Arthur, *Seek My Face, Speak My Name*, Jason Aronson, Northvale, New Jersey, 1992.

Grinspoon, Lester, and James Bakalar, *Psychedelic Drugs Reconsidered*, The Lindesmith Center, New York, 1997.

Grof, Stanislav, *Realms of the Human Unconscious*, Souvenir Press, London, 1979.

———, *LSD Psychotherapy*, Hunter House, Alameda, California, 1980.

———, *The Cosmic Game*, State University of New York Press, Albany, 1998.

———, *Psychology of the Future*, State University of New York Press, Albany, 2000.

Grof, Stanislav, and Christina Grof, editors, *Spiritual Emergency*, Jeremy Tarcher, Los Angeles, 1989.

Guthrie, Stewart, *Faces in the Clouds*, Oxford University Press, New York, 1993.

Happold, F. C., *Mysticism*, Penguin Books, New York, 1961.

Hofmann, Albert, *LSD: My Problem Child*, McGraw-Hill, New York, 1980.

Hooper, Judith, and Dick Teresi, *The Three-Pound Universe*, Jeremy Tarcher, Los Angeles, 1986.

Horgan, John, *The End of Science*, Broadway Books, New York, 1997.

———, *The Undiscovered Mind*, Free Press, New York, 1999.

Houston, Jean, *Jean Houston: A Mythic Life*, HarperSanFrancisco, 1996.

Huxley, Aldous, *The Perennial Philosophy*, Harper & Row, New York, 1944.

———, *Island*, Harper & Row, New York, 1962.

———, *The Doors of Perception*, Harper & Row, New York, 1990.

James, William, *The Varieties of Religious Experience*, Macmillan, New York, 1961.

Jansen, Karl, *Ketamine: Dreams and Realities*, MAPS, Sarasota, Florida, 2001.

Jonas, Hans, *The Gnostic Religion*, Beacon Press, Boston, 1991.

Kapleau, Philip, *The Three Pillars of Zen*, Anchor Books, Garden City, New York, 1980.

Katz, Steven, editor, *Mysticism and Philosophical Analysis*, Oxford University Press, New York, 1978.

Kramer, Joel, and Diana Alstad, *The Guru Papers*, Frog, Ltd., Berkeley, California, 1993.

LaBerge, Stephen, *Lucid Dreaming*, Ballantine Books, New York, 1985.

Lee, Martin, and Bruce Shlain, *Acid Dreams*, Grove Press, New York, 1992.

Lilly, John, *The Center of the Cyclone*, Bantam Books, New York, 1972.

———, *The Scientist*, Ronin Publishing, Berkeley, California, 1988.

Luna, Luis Eduardo, and Pablo Amaringo, *Ayahuasca Visions*, North Atlantic Books, Berkeley, California, 1991.

Marks, John, *The Search for the "Manchurian Candidate,"* W. W. Norton, New York, 1991.

Masters, Robert, and Jean Houston, *The Varieties of Psychedelic Experience,* Holt, Rinehart & Winston, New York, 1966.

Matthiessen, Peter, *The Snow Leopard,* Bantam Books, New York, 1978.

McGinn, Bernard, *The Foundations of Mysticism,* Crossroad, New York, 1991.

McKenna, Dennis, and Terence McKenna, *The Invisible Landscape,* Seabury Press, New York, 1975.

McKenna, Terence, *Food of the Gods,* Bantam Books, New York, 1992.

———, *True Hallucinations,* HarperSanFrancisco, 1993.

Narby, Jeremy, *The Cosmic Serpent,* Jeremy Tarcher/Putnam, New York, 1998.

Newberg, Andrew, and Eugene D'Aquili, *The Mystical Mind,* Fortress Press, Minneapolis, 1999.

Newberg, Andrew, Eugene D'Aquili, and Vince Rause, *Why God Won't Go Away,* Ballantine Books, New York, 2001.

Otto, Rudolf, *The Idea of the Holy,* Oxford University Press, New York, 1958.

Pagels, Elaine, *The Gnostic Gospels,* Vintage Books, New York, 1979.

Pellerin, Cheryl, *Trips,* Seven Stories Press, New York, 1998.

Persinger, Michael, *Neuropsychological Bases of God Beliefs,* Praeger, New York, 1987.

Ramachandran, V. S., and Sandra Blakeslee, *Phantoms in the Brain,* William Morrow, New York, 1998.

Ratsch, Christian, *Gateway to Inner Space: Sacred Plants, Mysticism, and Psychotherapy,* Avery Publishing Group, Garden City Park, New York, 1989.

Rawlinson, Andrew, *The Book of Enlightened Masters,* Open Court, Chicago, 1997.

Raymo, Chet, *Skeptics and True Believers,* Walker and Company, New York, 1998.

Salzman, Mark, *Lying Awake,* Knopf, New York, 2000.

Scholem, Gershom, *Major Trends in Jewish Mysticism,* Schocken Books, New York, 1946.

Schwartz, Tony, *What Really Matters,* Bantam Books, New York, 1995.

Shattuck, Roger, *Forbidden Knowledge,* Harcourt Brace, New York, 1996.

Shermer, Michael, *Why People Believe Weird Things,* W. H. Freeman, New York, 1997.

———, *How We Believe,* W. H. Freeman, New York, 2000.

Shulgin, Alexander, and Ann Shulgin, *PIHKAL,* Transform Press, Berkeley, California, 1991.

———, *TIHKAL,* Transform Press, Berkeley, California, 1997.

Siegel, Ronald, *Intoxication,* E. P. Dutton, New York, 1989.

Smith, Huston, *The World's Religions,* HarperSanFrancisco, 1991.

———, *Forgotten Truth,* HarperSanFrancisco, 1992.

———, *Cleansing the Doors of Perception,* Jeremy Tarcher/Putnam, New York, 2000.

———, *Why Religion Matters,* HarperCollins, New York, 2001.

Smoley, Richard, and Jay Kinney, *Hidden Wisdom,* Penguin/Arkana, New York, 1999.

Snyder, Solomon, *Drugs and the Brain,* Scientific American Books, New York, 1986.
Steindl-Rast, David, *A Listening Heart,* Crossroad, New York, 1999.
Stevens, Jay, *Storming Heaven,* Grove Press, New York, 1987.
Storr, Anthony, *Feet of Clay,* Free Press, New York, 1996.
Strassman, Rick, *DMT: The Spirit Molecule,* Park Street Press, Rochester, Vermont, 2001.
Tart, Charles, editor, *Altered States of Consciousness,* HarperCollins, New York, 1990.
Thornton, James, *A Field Guide to the Soul,* Bell Tower, New York, 1999.
Thurman, Robert, *Inner Revolution,* Riverhead Books, New York, 1998.
Walsh, Roger, and Francis Vaughan, editors, *Paths Beyond Ego,* Jeremy Tarcher/Putnam, New York, 1993.
Weinberg, Steven, *Dreams of a Final Theory,* Pantheon, New York, 1992.
———, *Facing Up,* Harvard University Press, Cambridge, 2001.
Wilber, Ken, *Grace and Grit,* Shambhala Publications, Boston, 1991.
———, *Sex, Ecology, Spirituality,* Shambhala Publications, Boston, 1995.
———, *A Brief History of Everything,* Shambhala Publications, Boston, 1996.
———, *Eye to Eye,* Shambhala Publications, Boston, 1996.
———, *The Marriage of Sense and Soul,* Broadway Books, New York, 1998.
———, *One Taste,* Shambhala Publications, Boston, 1999.
———, *Integral Psychology,* Shambhala Publications, Boston, 2000.
Wilson, Edward O., *Consilience,* Knopf, New York, 1998.
Wittgensein, Ludwig, *Tractatus Logico-Philosophicus,* Routledge, New York, 1990.
Zaehner, R. C., *Mysticism Sacred and Profane,* Oxford University Press, New York, 1961.

Index